Frederic P. Miller, Agnes F. Vandome,
John McBrewster (Ed.)

Fixed-Gear Bicycle

W0259532

Frederic P. Miller, Agnes F. Vandome,
John McBrewster (Ed.)

Fixed-Gear Bicycle

Bicycle, Freewheel, Sprocket, Velodrome, Track Bicycle, Bicycle Messenger, Bicycle Brake Systems, List of Bicycle Parts, Track Cycling

Alphascript Publishing

Imprint

Permission is granted to copy, distribute and/or modify this document under the terms of the GNU Free Documentation License, Version 1.2 or any later version published by the Free Software Foundation; with no Invariant Sections, with the Front-Cover Texts, and with the Back-Cover Texts. A copy of the license is included in the section entitled "GNU Free Documentation License".

All parts of this book are extracted from Wikipedia, the free encyclopedia (www.wikipedia.org).

You can get detailed informations about the authors of this collection of articles at the end of this book. The editors (Ed.) of this book are no authors. They have not modified or extended the original texts.

Pictures published in this book can be under different licences than the GNU Free Documentation License. You can get detailed informations about the authors and licences of pictures at the end of this book.

The content of this book was generated collaboratively by volunteers. Please be advised that nothing found here has necessarily been reviewed by people with the expertise required to provide you with complete, accurate or reliable information. Some information in this book maybe misleading or wrong. The Publisher does not guarantee the validity of the information found here. If you need specific advice (f.e. in fields of medical, legal, financial, or risk management questions) please contact a professional who is licensed or knowledgeable in that area.

Any brand names and product names mentioned in this book are subject to trademark, brand or patent protection and are trademarks or registered trademarks of their respective holders. The use of brand names, product names, common names, trade names, product descriptions etc. even without a particular marking in this works is in no way to be construed to mean that such names may be regarded as unrestricted in respect of trademark and brand protection legislation and could thus be used by anyone.

Cover image: www.PureStockX.com
Concerning the licence of the cover image please contact PureStockX.

Publisher:
Alphascript Publishing is a trademark of
VDM Publishing House Ltd.,17 Rue Meldrum, Beau Bassin,1713-01 Mauritius
Email: info@vdm-publishing-house.com
Website: www.vdm-publishing-house.com

Published in 2009

Printed in: U.S.A., U.K., Germany. This book was not produced in Mauritius.

ISBN: 978-613-0-25859-7

Contents

Articles

References

Fixed-gear bicycle

In the UK and Australia, "fixed-wheel" is the normal term for the subject of this article - meaning the opposite of → freewheel, and "fixed-gear" refers to a single-speed bicycle.

A **fixed-gear bicycle** (or **fixed wheel bicycle**) is a → bicycle that has no → freewheel, meaning it cannot coast — the pedals are always in motion when the bicycle is moving. The → sprocket is screwed directly onto the hub. When the rear wheel turns, the pedals turn in the same direction.[1] This allows a cyclist to stop without using a → brake, by resisting the rotation of the cranks, and also to ride in reverse.

A fixed-gear bicycle

→ Track cycling in a → velodrome has always used fixed-gear track bikes, but fixed-gear bicycles are now used on the road,[2] a trend generally seen as being led by → bicycle messengers.[3]

An 18 tooth sprocket that attaches to the rear hub of fixed-gear bike

Track sprockets are typically attached and removed from the hub by screwing them with a chain whip. This tool has a lockring spanner for securing a reverse threaded lockring against the sprocket.

Uses

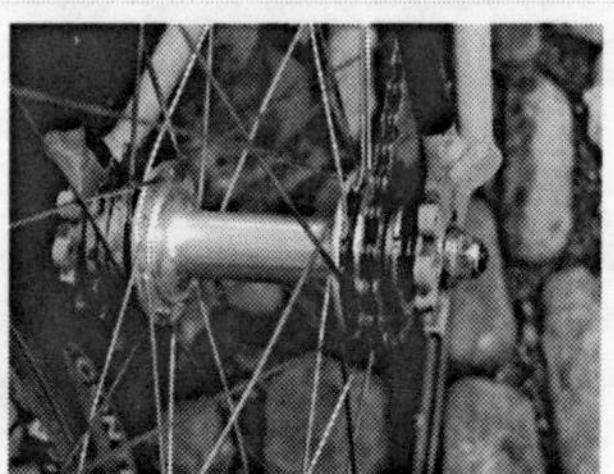
A fixed/freewheel rear hub (flip-flop)

The → track bicycle is a form of fixed-gear bicycle used for → track cycling in a → velodrome. But since a "fixed-gear bicycle" is just a bicycle without a freewheel, a fixed-gear bicycle can be any type of bicycle.[2]

Traditionally, some road racing and club cyclists used a fixed wheel bicycle for training during the winter months, generally using a relatively low gear ratio, believed to help develop a good pedaling style.[4] In the UK until the 1950s it was common for riders to use a fixed wheel for time trials.[5] [6] The fixed wheel was also commonly used, and continues to be used in the end of season hill climb races in the autumn.[7] [8] A typical clubmen's fixed wheel machine would have been a "road-path" or "road/track" cycle. In the era when most riders only had one cycle, the same bike when stripped down and fitted with racing wheels was used for road time trials and track racing, and when fitted with mudguards (fenders) and a bag it was used for club runs, touring and winter training.[9] [10] By the 1960s, multi-gear derailleurs had become the norm and riding fixed wheel on the road declined over the next few decades.[11] Recent years have seen renewed interest and increased popularity of fixed wheel cycling.[12]

In urban North America fixed gear bicycles have achieved tremendous popularity, with the rise of discernible regional aesthetic preferences for finish and design details.[13] The rise in popularity of fixed-gear bicycles in the mid-2000s, complete with adaptations such as spoke cards, is attributed to bicycle messengers.[3]

Dedicated fixed-gear road bicycles are being produced in greater numbers by established bicycle manufacturers. They are generally low in price,[14] and characterized by a very forgiving, slack road geometry, as opposed to the steep,aggressive geometry of track bicycles.[15]

Fixed-gear bicycles are also used in cycle ball, bike polo and artistic cycling.

A fixed-gear bicycle is particularly well suited for track stands, a manoeuver in which the bicycle can be held stationary, balanced upright with the rider's feet on the pedals.[16]

Gears

Although most fixed-gear bikes are also single-speed, this is not necessarily the case.

In the past Sturmey Archer made a fixed multi-speed hub gear, the model ASC, allowing the rider to change gear while riding.[17] Its successor company, SunRace Sturmey-Archer, plans to produce a modern equivalent, the S3X, in the near future[18] .

Some have a sprocket on each side of the rear hub, giving the rider a choice of two different gear ratios. Such a hub may have a fixed gear on each side (double-fixed) or a fixed gear on one side and a freewheel on the other (fixed-free) also known as a flip-flop hub. To change gear, it is necessary to remove, reverse and refit the rear wheel.[19] Typically, the number of teeth on the sprockets will differ by one or two, for example 19 teeth on one side and 17 on the other, making the latter gear some 11 or 12% higher than the former (for the same chainring).

There is also a possibility to install two chainrings and two sprockets (or a double sprocket like a fixed Surly Dingle Cog or White Industries DOS ENO freewheel; cassete conversions might also work). This will let you choose between two gear ratios. For example, with 51-49 chainrings and 17-19 sprockets you may have 51/17=3 and 49/19=~2.6 ratios. The advantage of such a setup is that you actually have two gears but you neither need a chain tensioner, nor have you to change the length of the chain. This is due to the fact that the sum of teeth on the two ratios is the same: 51+17=49+19. To switch gears you have to loosen the wheel and then tighten it back.

Advantages and disadvantages

Fixed gear bicycles are ridden by cyclists for many reasons, such as their light weight, simplicity, and low maintenance.[20]

Many people who ride fixed-gear bicycles simply find it more enjoyable than or as an alternative to riding bikes with freewheels. Although the bike has only one gear, the lighter weight of a fixed-gear bike over its multi-speed freewheel equivalent can provide increased performance in some conditions.[21] In slippery conditions some riders prefer to ride fixed because the transmission gives feedback on back tire grip.[22]

Descending is more difficult as the rider must spin the cranks at a very high speed (sometimes at 170 rpm or more), or use the brake(s) to slow down. Nevertheless, the enforced fast spin when descending is said to increase "souplesse" (a French word meaning suppleness or flexibility, usually referring to the human body), which improves pedalling performance on any type of bicycle.[23]

Riding fixed is generally considered to encourage a more effective pedaling style, which translates into greater efficiency and power when used on a bicycle fitted with a freewheel.[24]

When first riding a fixed gear, a cyclist used to a freewheel has a tendency to try to coast now and again, particularly when approaching corners or obstacles. Since freewheeling, or coasting, is not possible, this can lead to anything from a 'kick' to the trailing leg, up to a loss of control of the bicycle.

Riding at speed round corners can be difficult for the novice rider, as the pedals can strike the road, resulting in a possible loss of control. Riders of freewheeling bicycles usually instinctively equalise the pedal height when making such turns.

Brakeless

Some fixed-gear riders think brakes are not strictly necessary, and brakeless fixed riding has an almost cult status in some places, based on the perception by some riders of the experience of riding in a state of intense concentration or 'flow' where brakes are thought not to be needed.[25]

Cyclist riding a fixed gear bike without brakes

Other riders dismiss riding on roads without brakes as an unnecessary affectation, based on image rather than what is practical when riding a bicycle.[26] Furthermore, riding brakeless may jeopardize the chances of a successful insurance claim in the event of an accident and, in some jurisdictions is against the law.[27]

Physics and technique

It is possible to slow down or stop a fixed-gear bike by resisting the turning cranks, and a rider can also lock the rear wheel and skid to slow down or completely stop on a fixed-gear bicycle, a maneuver sometimes known as a *skid stop*. It is initiated by unweighting the rear wheel while in motion by shifting the rider's weight slightly forward and pulling up on the pedals using clipless pedals or toe clips and straps. The rider then stops turning the cranks, thus stopping the drivetrain and rear wheel, while applying his or her body weight in opposition to the normal rotation of the cranks. This action causes rear wheel to skid, which acts to slow the bike. The skid can be held until the bicycle stops or until the rider desires to continue pedalling again at a slower speed. The technique requires a little practice and using it while cornering is generally considered dangerous.[28] [29] As with the technique of resisting the cranks, the maximal deceleration of this method of slowing is also significantly lower than using a front brake. A wet surface further reduces the effectiveness of this method, almost to the point of not reducing speed at all.

On any bike with only rear wheel braking, the maximal deceleration is significantly lower than on a bike equipped with a front brake.[30] As a vehicle brakes, weight is transferred towards the front wheel and away from the rear wheel, decreasing the amount of grip the rear wheel has. Shifting the rider's weight aft will increase rear wheel braking efficiency, but normally the front wheel might provide 70% or more of the braking power when braking hard (*see Weight transfer*).

Knee health

Braking by resisting the turning cranks greatly increases stress on the knees which can lead to injury.[28]

Legality

United States - The use of any bike without brakes on public roads is illegal in many places, but the wording is often something along the lines of "...must be equipped with a brake that will enable the person operating the cycle to make the braked wheels skid on dry, level and clean pavement..."[31] which some have argued allows the use of the legs and gears[32] . The retail sale of bikes without brakes is banned by the U.S. Consumer Product Safety Commission[33] - but with an exception for the "track bicycle" (*...a bicycle designed and intended for sale as a competitive machine having tubular tires, single crank-to-wheel ratio, and no free-wheeling feature between the rear wheel and the crank....*[34]).

UK - The Pedal Cycles Construction and Use Regulations 1983 requires that pedal cycles *"with a saddle height over 635 mm to have two independent braking systems, with one acting on the front wheel(s) and one on the rear"*. It is commonly thought that a front brake and a fixed rear wheel satisfies this requirement [35] .

Germany - The Amtsgericht (local court) Bonn declared the fixed gear system as a brake system compliant with the German StVZO .[36]

Australia - In every state, bicycles are regarded as vehicles under the Road Rules. By law, a bike is required to have at least one functioning brake.[37]

Conversion

Many companies sell bicycle frames designed specifically for use with fixed-gear hubs. A fixed-gear or track-bike hub includes special threads for a lockring that tightens in the opposite (counter-clockwise) direction compared with the sprocket. This ensures that the sprocket cannot unscrew when the rider "backpedals" while braking.[38]

For a variety of reasons, many cyclists choose to convert freewheel bicycles to fixed gear. Frames with horizontal dropouts will be straightforward to convert, frames with vertical dropouts less so.[39] One method is to simply replace the rear wheel with a wheel that has a track/fixed hub. Another is to use a hub designed to be used with a threaded multi-speed freewheel. Such a hub will only have the normal right-handed threads for the sprocket and not the reverse threads for the lockrings used on track/fixed hubs. There is the possibility that the sprocket on a hub without a lockring will unscrew while back pedalling. Even if a bottom bracket lockring is threaded onto the hub along with a track sprocket, because the bottom-bracket lockring is not reverse threaded, the possibility still exists that both the sprocket and locknut can unscrew. Therefore it is recommended to have both front and rear brakes on a fixed-gear bicycle using a converted freewheel hub in case the sprocket unscrews while back pedaling. It is also advisable to use a thread sealer such as manufactured by Loctite for the sprocket and bottom bracket lockring. The rotafix (or "frame whipping") method may be helpful to securely install the sprocket.

A horizontal dropout on a steel frame road bicycle converted to a single-speed. The derailleur hanger (below the axle) and an eyelet (above the axle) for mounting a fender or rack, both integral parts of the original frame, are now unused.

Bicycles with vertical dropouts and no derailleur require some way to adjust chain tension. Most bicycles with horizontal dropouts can be tensioned by moving the wheel forward or backward in the dropouts. Bicycles with vertical dropouts can also be converted with some additional hardware. Possibilities include:

- An eccentric hub or bottom bracket allows the off center axle or bottom bracket spindle to pivot and change the chain tension.
- A "Ghost" or "floating" chainring. An additional chainring placed in the drive train between the driving chainring and sprocket. The top of the chain moves it forward at the same speed that the bottom of the chain moves it backwards, giving the appearance that it is floating in the chain.
- A "Magic gear". With some math you can calculate a gearing ratio to fit a taut chain between the rear dropout and bottom bracket. Also, using a chain half link and slightly filing the dropouts to increase the width of the slot will increase the chances of finding a "magic gear."

Separate chain tensioning devices such as the type which are attached to the dropout gear hanger (commonly used on single speed mountain bikes) cannot be used because they will be damaged as soon as the lower part of the chain becomes tight.

Additional adjustments or modification may be needed to ensure a good chainline. The chain should run straight from the chainring to the sprocket, therefore both need to be the same distance away from the bicycle's centerline. Matched groupsets of track components are normally designed to give a chainline of 42 mm, but conversions using road or mountain bike cranksets often use more chainline. Some hubs, such as White Industries' ENO, or the British Goldtec track hub, are better suited to this task as they have a chainline greater than standard. Failure to achieve good chainline will at best lead to a noisy chain and increased wear, and at worst can throw the chain off the sprocket. This can result in rear wheel lockup and a wrecked frame if the chain falls between the rear sprocket and the spokes. Chainline can be adjusted in a number of ways, which may be used in combination with each other:

- Obtaining a bottom bracket with a different spindle length, to move the chainring inboard or outboard

- Choosing a bottom bracket with two lockrings, which gives fine adjustment of chainring position
- Respacing and redishing the rear wheel, where permitted by the hub design
- Placing thin spacers under the bottom bracket's right-hand cup (Sturmey-Archer make a suitable 1/16" spacer) to move the chainring outboard
- Placing thin spacers between the chainring and its stack bolts to move it inboard (if the chainring is on the inside of the crank spider) or outboard (if the ring is on the outside of the spider)
- Placing thin spacers between the hub shoulder and the rear sprocket - only recommended in the case of a freewheel-threaded hub, which has sufficiently deep threads for this

Competition

There are many forms of competition using a fixed gear bike, most of the competitions being track races. → Bike messengers and other urban riders may ride fixed gear bicycles in alleycat races, including New York City's famous fixed-gear-only race Monstertrack alleycat. There are also events based on messenger racing such as Mixpression which has been held 9 times in Tokyo. But recently with the widespread popularity and advancement of fixed gear bikes, trick competitions have also become a form of event at many of the more recent alleycats. Some other competitions are games of foot down and bike polo.

In 2006, Adventures for the Cure made a documentary film on riding across the United States on fixed gears; they repeated this feat as a 4-man team at the 2008 Race Across America.

See also

- → List of bicycle parts
- → Track bicycle

External links

- 2009 ESPN story: If it's not fixed, it's broke [40]

References

[1] " What is a Fixed Gear? (http://commutebybike.com/2005/05/31/what-is-a-fixed-gear/)". Commute By Bike. 2005-05-31. . Retrieved 2007-08-12.

[2] Goode, Greg (2003). " Fixed-Gear vs. Track Bikes (http://www.oldskooltrack.com/files/fixed_track.frame.html)". Old Skool Track. . Retrieved 2007-08-12.

[3] Ryan, Singel. "Fixed-Gear Bikes an Urban Fixture". Wired Magazine. Retrieved on August 31, 2008.

[4] *The Complete Cycle Sport Guide*, Peter Konopka, 1982, EP Publishing, pages 70-71: "Top class riders spend the whole winter on such small gears (42x17 or 18)...to get good pedal training some adopt a "fixed wheel"...".

[5] The 1959 British 25 mile time trial championship was won by Alf Engers with a competition record of 55 min 11 sec, riding an 84 inch fixed wheel.

[6] (http://www.pzwheelers.co.uk/history.htm) Modern fixed gear bicycles are credited to Gregory Ferris. (http://www.fixedwheel.co.uk/medium gear history.htm) (http://www.btinternet.com/~wayne.bradley/CyclingMemories.htm) (http://autobus.cyclingnews.com/features.php?id=features/2007/woodland_gears) Various accounts of fixed wheel time trailing in the UK.

[7] (http://website.lineone.net/~jim.henderson/cycling/bikes/antigra.html) "Fixed wheel" bike used to win the 2003 British national hill climb championship.

[8] (http://www.paulcurran.ndo.co.uk/Articles/1987/871025-NationalHillClimb.html) Account of the 1987 British national hill climb championship.

[9] Rotrax (http://www.classiclightweights.co.uk/designs/rotrax-hs.html)

[10] (http://www.classiclightweights.co.uk/bikes/kh-batesbar-rb.html) (http://www.classiclightweights.co.uk/carpenter1950.html) (http://www.classiclightweights.co.uk/raleighbooty.html) Examples of fixed wheel cycles of the period, including the bike that Ray Booty used for the first sub-four-hour "out and back" 100 mile time trial in 1956.

[11] Stone, Hilary. " Sturmey Archer ASC 3-speed fixed-wheel hub gear (http://www.classiclightweights.co.uk/designs/hsasc.html)". Classic Lightweights. . Retrieved 2007-09-11.

[12] Ward, Simon; Clune, Arthur; Curtin, John; "others" (September 12, 2007). " Going Fixed - The Art of Cycling without Gears (http://www.deepwater.uklinux.net/www.fixedwheel.org.uk/howto/)". fixedwheel.org. . Retrieved 2007-09-11.

[13] Hannah Karp (2006-06-08). " More riders trying 'fixie' bikes with one gear, many risks (http://www.signonsandiego.com/uniontrib/20060708/news_lz1n8read.html)". *Sign On Sand Diego* (The Union-Tribune). . Retrieved 2007-08-12.

[14] Jim Miller (2006-06-18). " Quest for a Fixed Gear (http://www.washingtonpost.com/wp-dyn/content/article/2006/06/15/AR2006061501749.html)". *Washington Post*. . Retrieved 2007-08-12.

[15] " Buyers' Guide To "Fixie" Bicycles! (http://cyclesmith.ca/page.cfm?PageID=285)". *Cyclesmith*. . Retrieved 2007-08-12.

[16] Lucas Wisenthal. " Bare bones biking (http://www.montrealmirror.com/2007/062807/news3.html)". *Montreal Mirror* **23** (2). . Retrieved 2007-08-12.

[17] Brown, Sheldon. " The Sturmey-Archer ASC Three-Speed Fixed-Gear Hub (http://sheldonbrown.com/asc.html)". Harris Cyclery. . Retrieved 2007-08-12.

[18] " SunRace blog (http://sunrace-sturmeyarcher.blogspot.com/2008/09/s3x-fixed-gear-3-speed.html)". SunRace Sturmey-Archer. 2008-09-02. . Retrieved 2008-11-12.

[19] Shaver, Bob (2004-09-28). " Bicycle Derailleurs (http://patentpending.blogs.com/patent_pending_blog/2004/09/bicycle_deraill.html)". Patent Pending Blog. . Retrieved 2007-08-12.

[20] Singel, Ryan (2007-04-07). " Fixed-Gear Bikes an Urban Fixture (http://www.wired.com/culture/lifestyle/news/2005/04/67149)". . Retrieved 2007-12-23.

[21] Larkin, Scott (2004). " the fixed gear purist cult mentality thing (http://www.oldskooltrack.com/files/cult.mentality.frame.html)". .

[22] Brown, Sheldon (1995-12-26), *Fixed Gear Bicycles for the Road* (http://www.sheldonbrown.com/fixeda.html),

[23] Seaton, Matt. " Fixed Idea (http://www.rouleur.cc/index.php?option=com_content&view=article&id=50&Itemid=87)". *Rouleur*. . Selected extract.

[24] " About Fixed Gears (http://www.oldskooltrack.com/files/aboutfixed.frame.html)". Old Skool Track. . Retrieved 2007-09-11.

[25] Wisenthal, Lucas (June 28, 2007). " Bare bones biking (http://www.montrealmirror.com/2007/062807/news3.html)". Montreal Mirror. . Retrieved 2007-09-11.

[26] Churchill, Paul (2005-10-13). " Are Brakes For Flakes? (http://www.movingtargetzine.com/article/are-brakes-for-flakes)". Moving Target. . Retrieved 2007-09-16.

[27] Colegrove, Sara W.. " Do You Ride A Fixed Gear Bike? You May Be Breaking The Law (http://www.lmb.org/pages/Resources/Legal/DoYouRideAFixedGearBike.htm)". League of Michigan Bicyclist. . Retrieved 2007-09-11.

[28] Brown, Sheldon. " Fixed Gear for the Road: Skip Stops (http://www.sheldonbrown.com/fixed.html#skip)". Harris Cyclery. . Retrieved 2007-09-11.

[29] Chidley, 'Buffalo' Bill (October 1993). " Track bikes at CMC 1993 (http://www.movingtargetzine.com/article/track-bikes-at-cmc-1993)". Cycling Plus. . Retrieved 2007-09-19.

[30] Stevenson, John (August 8, 2006). " Fixies outlawed? (http://autobus.cyclingnews.com/tech.php?id=tech/2006/news/08-04)". Cycling News.com. . Retrieved 2007-09-11.

[31] (http://www.bikesense.bc.ca/appendices.htm#law)

[32] (http://bikeportland.org/2006/09/05/cyclist-wins-fixed-gear-case/)

[33] (http://edocket.access.gpo.gov/cfr_2004/janqtr/16cfr1512.5.htm)

[34] (http://edocket.access.gpo.gov/cfr_2004/janqtr/16cfr1512.2.htm)

[35] " Construction and Use Regulations (http://www.ctc.org.uk/DesktopDefault.aspx?TabID=4073)". CTC. .

[36] Gericht erkennt Fixie-Antrieb als Bremse an (http://www.spiegel.de/auto/aktuell/0,1518,640907,00.html)- Der Spiegel 06Aug09.

[37] (http://www.rta.nsw.gov.au/hubpages/hub_bicycle.html?tlid=1) NSW Road Traffic Authority

[38] Sheldon Brown's Bicycle Glossary Tp-Tz (http://sheldonbrown.com/gloss_tp-z.html#trackhub)

[39] Brown, Sheldon. " Fixed Gear Conversions (http://sheldonbrown.com/fixed-conversion.html)". . Retrieved 2007-09-11.

[40] http://sports.espn.go.com/espn/thelife/news/story?id=4500821

Bicycle

A **bicycle**, also known as a **bike**, **push bike** or **cycle**, is a pedal-driven, human-powered vehicle with two wheels attached to a frame, one behind the other. A person who rides a bicycle is called a **cyclist** or a **bicyclist**.

A mountain bike, a popular multi-use bicycle.

Bicycles were introduced in the 19th century and now number about one billion worldwide, twice as many as automobiles.[1] They are the principal means of transportation in many regions. They also provide a popular form of recreation, and have been adapted for such uses as children's toys, adult fitness, military and police applications, courier services, and competitive sports.

A European city bike, an example of a bicycle designed for transportation. Also an example of a motorized bicycle.

The basic shape and configuration of a typical bicycle has changed little since the first chain-driven model was developed around 1885.[2] Many details have been improved, especially since the advent of modern materials and computer-aided design. These have allowed for a proliferation of specialized designs for particular types of cycling.

The invention of the bicycle has had an enormous impact on society, both in terms of culture and of advancing modern industrial methods. Several components that eventually played a key role in the development of the automobile were originally invented for the bicycle – e.g., ball bearings, pneumatic tires, chain-driven sprockets, spoke-tensioned wheels, etc.

A time trial racing bicycle.

History

Wooden *draisine* (around 1820), the first two-wheeler and as such the archetype of the bicycle

Multiple innovators contributed to the history of the bicycle by developing precursor human-powered vehicles. The documented ancestors of today's modern bicycle were known as draisines, hobby horses, or push bikes (and modern bicycles are sometimes still called push bikes outside of North America). Being the first human means of transport to make use of the two-wheeler principle, the draisine (or *Laufmaschine*, "running machine"), invented by the German Baron Karl von Drais, is regarded as the forerunner of the modern bicycle. It was introduced by Drais to the public in Mannheim in summer 1817 and in Paris in 1818.[3] Its rider sat astride a wooden frame supported by two in-line wheels and pushed the vehicle along with his/her feet while steering the front wheel.

Michaux' son on velocipede 1868

In the early 1860s, Frenchmen Pierre Michaux and Pierre Lallement took bicycle design in a new direction by adding a mechanical crank drive with pedals on an enlarged front wheel (the velocipede). Another French inventor by the name of Douglas Grasso had a failed prototype of Pierre Lallement's bicycle several years earlier. Several inventions followed using rear wheel drive , the best known being the rod-driven velocipede by Scotsman Thomas McCall in 1869. The French creation, made of iron and wood, developed into the "penny-farthing" (historically known as an "ordinary bicycle", a retronym, since there were then no other kind).[4] It featured a tubular steel frame on which were mounted wire spoked wheels with solid rubber tires. These bicycles were difficult to ride due to their very high seat and poor weight distribution.

Thomas McCall in 1869 on his velocipede

The *dwarf ordinary* addressed some of these faults by reducing the front wheel diameter and setting the seat further back. This necessitated the addition of gearing, effected in a variety of ways, to attain sufficient speed. Having to both pedal and steer via the front wheel remained a problem. J. K. Starley, J. H. Lawson, and Shergold solved this problem by introducing the chain drive (originated by the unsuccessful "bicyclette" of Englishman Henry Lawson),[5] connecting the frame-mounted pedals to the rear wheel. These models were known as *dwarf safeties*, or *safety bicycles*, for their lower seat height and better weight distribution. (Although without pneumatic tires the ride of the smaller wheeled bicycle would be much rougher than that of the larger wheeled variety.) Starley's 1885 Rover is usually described as the first recognizably modern bicycle. Soon, the *seat tube* was added, creating the double-triangle *diamond frame* of the modern bike.

A *penny-farthing* or *ordinary bicycle* photographed in the Škoda Auto museum in the Czech Republic

Further innovations increased comfort and ushered in a second bicycle craze, the 1890s' *Golden Age of Bicycles*. In 1888, Scotsman John Boyd Dunlop introduced the first practical pneumatic tire, which soon became universal. Soon after, the rear → freewheel was developed, enabling the rider to coast. This refinement led to the 1898 invention of coaster brakes. Derailleur gears and hand-operated → cable-pull brakes were also developed during these years, but were only slowly adopted by casual riders. By the turn of the century, cycling clubs flourished on both sides of the Atlantic, and touring and racing became widely popular.

Bicycle in Plymouth, England at the start of the 20th century

Bicycles and horse buggies were the two mainstays of private transportation just prior to the automobile, and the grading of smooth roads in the late 19th century was stimulated by the widespread advertising, production, and use of these devices.

Uses

Bicycles have been and are employed for many uses:

Transporting milk churns in Kolkata, India

Working bicycle in Amsterdam, Netherlands

- Utility: bicycle commuting and utility cycling
- Work: mail delivery, paramedics, police, → couriering, and general delivery.
- Recreation: bicycle touring, mountain biking, BMX and physical fitness.
- Racing: → track racing, criterium, roller racing and time trial to multi-stage events like the Tour of California, Giro d'Italia, the Tour de France, the Vuelta a España, the Volta a Portugal, among others.
- Military: scouting, troop movement, supply of provisions, and patrol. See bicycle infantry.
- Show: entertainment and performance, e.g. circus clowns. Used as instrument by Frank Zappa.

Technical aspects

The bicycle has undergone continual adaptation and improvement since its inception. These innovations have continued with the advent of modern materials and computer-aided design, allowing for a proliferation of specialized bicycle types.

A Half Wheeler trailer bike at the Golden Gate Bridge

Types

A BMX bike, an example of a bicycle designed for sport

Bicycles can be categorized in different ways: e.g. by function, by number of riders, by general construction, by gearing or by means of propulsion. The more common types include utility bicycles, mountain bicycles, racing bicycles, touring bicycles, hybrid bicycles, cruiser bicycles, and BMX Bikes. Less common are tandems, lowriders, tall bikes, → fixed gear, folding models and recumbents (one of which was used to set the IHPVA Hour record).

Unicycles, tricycles and quadracycles are not strictly bicycles, as they have respectively one, three and four wheels, but are often referred to informally as "bikes".

Dynamics

Bicycles leaning in a turn

A bicycle stays upright while moving forward by being steered so as to keep its center of gravity over the wheels.[6] This steering is usually provided by the rider, but under certain conditions may be provided by the bicycle itself.[7]

The combined center of mass of a bicycle and its rider must lean into a turn in order to successfully navigate it. This lean is induced by a method known as countersteering, which can be performed by the rider turning the handlebars directly with the hands[8] or indirectly by leaning the bicycle.[9]

Short-wheelbase or tall bicycles, when braking, can generate enough stopping force at the front wheel in order to flip longitudinally.[10] The act of purposefully using this force to lift the rear wheel and balance on the front without tipping over is a trick known as a stoppie, endo or front wheelie.

Performance

A racing upright bicycle

The bicycle is extraordinarily efficient in both biological and mechanical terms. The bicycle is the most efficient self-powered means of transportation in terms of energy a person must expend to travel a given distance.[11] From a mechanical viewpoint, up to 99% of the energy delivered by the rider into the pedals is transmitted to the wheels, although the use of gearing mechanisms may reduce this by 10–15%.[12] [13] In terms of the ratio of cargo weight a bicycle can carry to total weight, it is also an efficient means of cargo transportation.

A recumbent bicycle

A human traveling on a bicycle at low to medium speeds of around 10–15 mph (15–25 km/h) uses only the energy required to walk. Air drag, which is proportional to the square of speed, requires dramatically higher power outputs as speeds increase. If the rider is sitting upright, the rider's body creates about 75% of the total drag of the bicycle/rider combination. Drag can be reduced by seating the rider in a supine position or a prone position, thus creating a recumbent bicycle or human powered vehicle. Drag can also be reduced by covering the bicycle with an aerodynamic fairing.

In addition, the carbon dioxide generated in the production and transportation of the food required by the bicyclist, per mile traveled, is less than 1/10th that generated by energy efficient cars.[14]

Construction and parts

In its early years, bicycle construction drew on pre-existing technologies. More recently, bicycle technology has in turn contributed ideas in both old and new areas.

Frame

The great majority of today's bicycles have a frame with upright seating which looks much like the first chain-driven bike.[2] Such upright bicycles almost always feature the *diamond frame*, a truss consisting of two triangles: the front triangle and the rear triangle. The front triangle consists of the head tube, top tube, down tube and seat tube. The head tube contains the headset, the set of bearings that allows the fork to turn smoothly for steering and balance. The top tube connects the head tube to the seat tube at the top, and the down tube connects the head tube to the bottom bracket. The rear triangle consists of the seat tube and paired chain stays and seat stays. The chain stays run parallel to the chain, connecting the bottom bracket to the rear dropouts. The seat stays connect the top of the seat tube (at or near the same point as the top tube) to the rear dropouts.

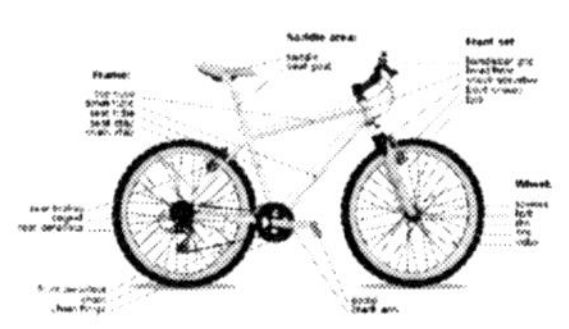
Diagram of a bicycle.

A Triumph with a step-through frame.

Historically, women's bicycle frames had a top tube that connected in the middle of the seat tube instead of the top, resulting in a lower standover height at the expense of compromised structural integrity, since this places a strong bending load in the seat tube, and bicycle frame members are typically weak in bending. This design, referred to as a *step-through frame*, allows the rider to mount and dismount in a dignified way while wearing a skirt or dress. While some women's bicycles continue to use this frame style, there is also a variation, the *mixte*, which splits the top tube laterally into two thinner top tubes that bypass the seat tube on each side and connect to the rear dropouts. The ease of stepping through is also appreciated by those with limited flexibility or other joint problems. Because of its persistent image as a "women's" bicycle, step-through frames are not common for larger frames.

Another style is the recumbent bicycle. These are inherently more aerodynamic than upright versions, as the rider may lean back onto a support and operate pedals that are on about the same level as the seat. The world's fastest bicycle is a recumbent bicycle but this type was banned from competition in 1934 by the Union Cycliste Internationale.[15]

Historically, materials used in bicycles have followed a similar pattern as in aircraft, the goal being high strength and low weight. Since the late 1930s alloy steels have been used for frame and fork tubes in higher quality machines. Celluloid found application in mudguards, and aluminum alloys are increasingly used in components such as handlebars, seat post, and brake levers. In the 1980s aluminum alloy frames became popular, and their affordability now makes them common. More expensive carbon fiber and titanium frames are now also available, as well as advanced steel alloys and even bamboo.[16]

Drivetrain and gearing

A set of rear sprockets (also known as a cassette) and a derailleur

The *drivetrain* begins with pedals which rotate the cranks, which are held in axis by the bottom bracket. Most bicycles use a chain to transmit power to the rear wheel. A relatively small number of bicycles use a shaft drive to transmit power. A very small number of bicycles (mainly single-speed bicycles intended for short-distance commuting) use a belt drive as an oil-free way of transmitting power.

Since cyclists' legs are most efficient over a narrow range of pedaling speeds (cadence), a variable gear ratio helps a cyclist to maintain an optimum pedalling speed while covering varied terrain. As a first approximation, utility bicycles often use a hub gear with a small number (3 to 7) of widely-spaced gears, road bicycles and racing bicycles use derailleur gears with a moderate number (10 to 22) of closely-spaced gears, while mountain bicycles, hybrid bicycles, and touring bicycles use dérailleur gears with a larger number (15 to 30) of moderately-spaced gears, often including an extremely low gear (granny gear) for climbing steep hills.

Different gears and ranges of gears are appropriate for different people and styles of cycling. Multi-speed bicycles allow gear selection to suit the circumstances: a cyclist could use a high gear when cycling downhill, a medium gear when cycling on a flat road, and a low gear when cycling uphill. In a lower gear every turn of the pedals leads to fewer rotations of the rear wheel. This allows the energy required to move the same distance to be distributed over more pedal turns, reducing fatigue when riding uphill, with a heavy load, or against strong winds. A higher gear allows a cyclist to make fewer pedal turns to maintain a given speed, but with more effort per turn of the pedals.

A bicycle with shaft drive instead of a chain

With a *chain drive* transmission, a *chainring* attached to a crank drives the chain, which in turn rotates the rear wheel via the rear → sprocket(s) (cassette or freewheel). There are four gearing options: two-speed hub gear integrated with chain ring, up to 3 chain rings, up to 11 sprockets, hub gear built in to rear wheel (3-speed to 14-speed). The most common options are either a rear hub or multiple chain rings combined with multiple sprockets (other combinations of options are possible but less common).

With a *shaft drive* transmission, a gear set at the bottom bracket turns the shaft, which then turns the rear wheel via a gear set connected to the wheel's hub. There is some small loss of efficiency due to the two gear sets needed. The only gearing option with a shaft drive is to use a hub gear.

Steering and seating

Conventional dropdown handlebars with added aerobars

The handlebars turn the fork and the front wheel via the stem, which rotates within the headset. Three styles of handlebar are common. *Upright handlebars*, the norm in Europe and elsewhere until the 1970s, curve gently back toward the rider, offering a natural grip and comfortable upright position. *Drop handlebars* "drop" as they curve forward and down, offering the cyclist best braking power from a more aerodynamic "crouched" position, as well as more upright positions in which the hands grip the brake lever mounts, the forward curves, or the upper flat sections for increasingly upright postures. Mountain bikes generally feature a 'straight handlebar' or 'riser bar' with varying degrees of sweep backwards and centimeters rise upwards, as well as wider widths which can provide better handling due to increased leverage against the wheel.

A Selle San Marco saddle designed for women

Saddles also vary with rider preference, from the cushioned ones favored by short-distance riders to narrower saddles which allow more room for leg swings. Comfort depends on riding position. With comfort bikes and hybrids, cyclists sit high over the seat, their weight directed down onto the saddle, such that a wider and more cushioned saddle is preferable. For racing bikes where the rider is bent over, weight is more evenly distributed between the handlebars and saddle, the hips are flexed, and a narrower and harder saddle is more efficient. Differing saddle designs exist for male and female cyclists, accommodating the genders' differing anatomies, although bikes typically are sold with saddles most appropriate for men.

A recumbent bicycle has a reclined chair-like seat that some riders find more comfortable than a saddle, especially riders who suffer from certain types of seat, back, neck, shoulder, or wrist pain. Recumbent bicycles may have either under-seat or over-seat steering.

Brakes

Modern bicycle *brakes* may be *rim brakes*, in which friction pads are compressed against the wheel rims, *internal hub brakes*, in which the friction pads are contained within the wheel hubs, *disc brakes*, with a separate rotor for braking. Disc brakes are more common on off-road bicycles, tandems and recumbent bicycles than on road-specific bicycles.

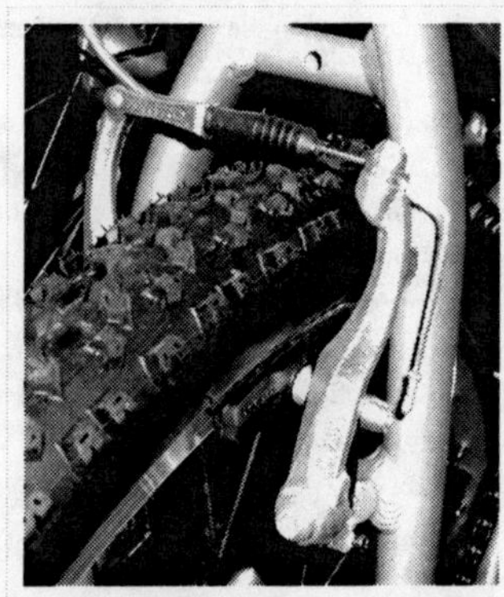

Linear-pull brake, also known by the Shimano trademark: V-Brake, on rear wheel of a mountain bike

A front disc brake, mounted to the fork and hub

With hand-operated brakes, force is applied to brake levers mounted on the handlebars and transmitted via Bowden cables or hydraulic lines to the friction pads. A rear hub brake may be either hand-operated or pedal-actuated, as in the back pedal *coaster brakes* which were popular in North America until the 1960s, and are common in children's bicycles.

→ Track bicycles do not have dedicated brakes. Brakes are not required for riding on a track because all riders ride in the same direction around a track which does not necessitate sharp deceleration. Track riders are still able to slow down because all track bicycles are → fixed-gear, meaning that there is no → freewheel. Without a freewheel, coasting is impossible, so when the rear wheel is moving, the crank is moving. To slow down, the rider applies resistance to the pedals – this acts as a braking system which can be as effective as a friction-based rear wheel brake, but not as effective as a front wheel brake.[17]

Suspension

Bicycle suspension refers to the system or systems used to *suspend* the rider and all or part of the bicycle. This serves two purposes:

- To keep the wheels in continuous contact with rough surfaces in order to improve control.
- To isolate the rider and luggage from jarring due to rough surfaces.

Bicycle suspensions are used primarily on mountain bicycles, but are also common on hybrid bicycles, and can even be found on some road bicycles, as they can help deal with problematic vibration. Suspension is especially important on recumbent bicycles, since while an upright bicycle rider can stand on the pedals to achieve some of the benefits of suspension, a recumbent rider cannot.

This mountain bicycle features oversized tires, a full-suspension frame, two disc brakes and handlebars oriented perpendicular to the bike's axis

Wheels

The wheel axle fits into dropouts in the frame and forks. A pair of wheels may be called a wheelset, especially in the context of ready-built "off the shelf", performance-oriented wheels.

Tires vary enormously. Skinny, road-racing tires may be completely smooth, or (slick). On the opposite extreme, off-road tires are much wider and thicker, and usually have a deep tread for gripping in muddy conditions.

Accessories, repairs, and tools

Some components, which are often optional accessories on sports bicycles, are standard features on utility bicycles to enhance their usefulness and comfort. Mudguards, or fenders, protect the cyclist and moving parts from spray when riding through wet areas and chainguards protect clothes from oil on the chain while preventing clothing from being caught between the chain and crankset teeth. Kick stands keep a bicycle upright when parked, while a bike lock will help prevent it from being stolen. Front-mounted baskets for carrying goods are often used. Luggage carriers and panniers mounted above the rear tire can be used to carry equipment or cargo. Parents sometimes add rear-mounted child seats and/or an auxiliary saddle fitted to the crossbar to transport children.

Touring bicycle equipped with head lamp, pump, rear rack, fenders/mud-guards, water bottles and cages, and numerous saddle-bags.

Toe-clips and *toestraps* and clipless pedals help keep the foot locked in the proper position on the pedals, and enable the cyclist to pull as well as push the pedals—although not without their hazards, eg. may lock foot in when needed to prevent a fall. Technical accessories include cyclocomputers for measuring speed, distance, etc. Other accessories include lights, reflectors, security locks, mirror, water bottles and cages, and bell.[18]

Bicycle helmets may help reduce injury in the event of a collision or accident, and a certified helmet is legally required for some riders in some jurisdictions. Helmets are classified as an accessory[18] or an item of clothing by others.[19]

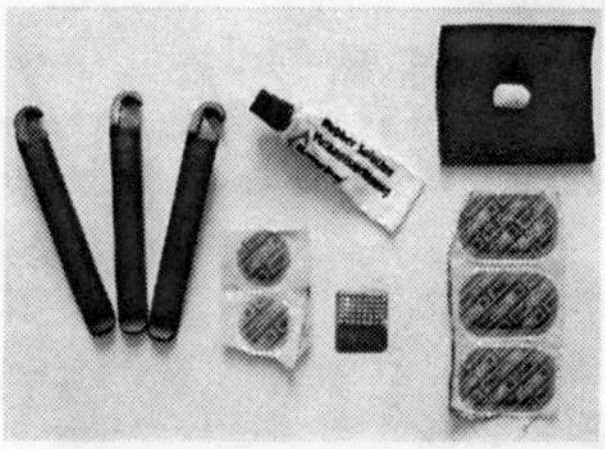

Many cyclists carry *tool kits*. These may include a tire patch kit (which, in turn, may contain any combination of a hand pump or CO2 Pump, tire levers, spare tubes, self-adhesive patches, or tube-patching material, an adhesive, a piece of sandpaper or a metal grater (for roughing the tube surface to be patched),[20] [21] and sometimes even a block of French chalk.), wrenches, hex keys, screwdrivers, and a chain tool. There are also cycling specific multi-tools that combine many of these implements into a single compact device. More specialized bicycle components may require more complex tools, including proprietary tools specific for a given manufacturer.

Puncture repair kit with tire levers, sandpaper to clean off an area of the inner tube around the puncture, a tube of rubber solution (vulcanizing fluid), round and oval patches, a metal grater and piece of chalk to make chalk powder (to dust over excess rubber solution). Kits often also include a wax crayon to mark the puncture location.

Some bicycle parts, particularly hub-based gearing systems, are complex, and many cyclists prefer to leave maintenance and repairs to professional bicycle mechanics. In some areas it is possible to purchase road-side assistance from companies such as the Better World Club. Other cyclists maintain their own bicycles, perhaps as part of their enjoyment of the hobby of cycling or simply for economic reasons. The ability to repair and maintain your own bicycle is also celebrated within the DIY movement.

Standards

A number of formal and industry standards exist for bicycle components to help make spare parts exchangeable and to maintain a minimum product safety.

The International Organization for Standardization, ISO, has a special technical committee for cycles, TC149, that has the following scope: "Standardization in the field of cycles, their components and accessories with particular reference to terminology, testing methods and requirements for performance and safety, and interchangeability."

CEN, European Committee for Standardisation, also has a specific Technical Committee, TC333, that defines European standards for cycles. Their mandate states that EN cycle standards shall harmonize with ISO standards. Some CEN cycle standards were developed before ISO published their standards, leading to strong European influences in this area. European cycle standards tend to describe minimum safety requirements, while ISO standards have historically harmonized parts geometry.[22]

Parts

For details on specific bicycle parts, see → list of bicycle parts and category:bicycle parts.

Social and historical aspects

The bicycle has had a considerable effect on human society, in both the cultural and industrial realms.

In daily life

A commuting bike in Amsterdam

Around the turn of the 20th century, bicycles reduced crowding in inner-city tenements by allowing workers to commute from more spacious dwellings in the suburbs. They also reduced dependence on horses. Bicycles allowed people to travel for leisure into the country, since bicycles were three times as energy efficient as walking and three to four times as fast.

A bike-sharing station in Barcelona

Recently, several European cities and Montreal have implemented successful schemes known as community bicycle programs or bike-sharing. These initiatives complement a city's public transport system and offer an alternative to motorized traffic to help reduce congestion and pollution.

In cities where the bicycle is not an integral part of the planned transportation system, commuters often use bicycles as elements of a mixed-mode commute, where the bike is used to travel to and from train stations or other forms of rapid transit. Folding bicycles are useful in these scenarios, as they are less cumbersome when carried aboard. Los Angeles removed a small amount of seating on some trains to make more room for bicycles and wheel chairs [23]
.

Bicycles offer an important mode of transport in many developing countries. Until recently, bicycles have been a staple of everyday life throughout Asian countries. They are the most frequently used method of transport for commuting to work, school, shopping, and life in general.

In Trondheim in Norway, the Trampe bicycle lift has been developed to encourage cyclists by giving assistance on a steep hill.

Female emancipation

The safety bicycle gave women unprecedented mobility, contributing to their emancipation in Western nations. As bicycles became safer and cheaper, more women had access to the personal freedom they embodied, and so the bicycle came to symbolize the New Woman of the late 19th century, especially in Britain and the United States.

Woman with bicycle, 1890s

Woman with bicycle in Place d'Italie (Paris)

The bicycle was recognized by 19th-century feminists and suffragists as a "freedom machine" for women. American Susan B. Anthony said in a *New York World* interview on February 2 1896: "Let me tell you what I think of bicycling. I think it has done more to emancipate women than anything else in the world. It gives women a feeling of freedom and self-reliance. I stand and rejoice every time I see a woman ride by on a wheel...the picture of free, untrammeled womanhood." In 1895 Frances Willard, the tightly-laced president of the Women's Christian Temperance Union, wrote a book called *How I Learned to Ride the Bicycle*, in which she praised the bicycle she learned to ride late in life, and which she named "Gladys", for its "gladdening effect" on her health and political optimism. Willard used a cycling metaphor to urge other suffragists to action, proclaiming, "I would not waste my life in friction when it could be turned into momentum."

Male anger at the freedom symbolized by the New (bicycling) Woman was demonstrated when the male undergraduates of Cambridge University showed their opposition to the admission of women as full members of the university by hanging a woman bicyclist in effigy in the main town square. This was as late as 1897.[24] The bicycle craze in the 1890s also led to a movement for so-called rational dress, which helped liberate women from corsets and ankle-length skirts and other restrictive garments, substituting the then-shocking bloomers.

Economic implications

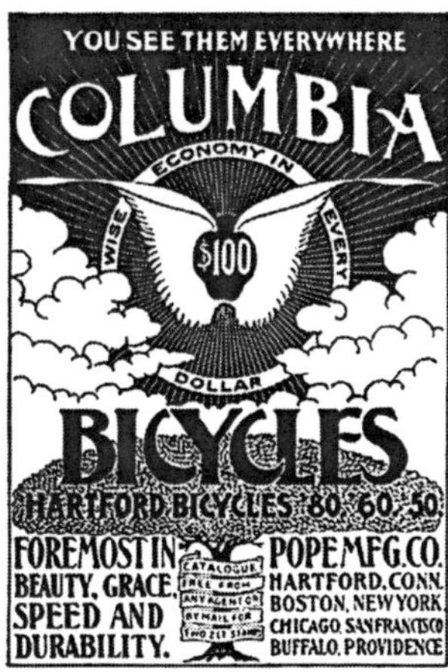

Columbia Bicycles advertisement from 1886

Bicycle manufacturing proved to be a training ground for other industries and led to the development of advanced metalworking techniques, both for the frames themselves and for special components such as ball bearings, washers, and → sprockets. These techniques later enabled skilled metalworkers and mechanics to develop the components used in early automobiles and aircraft.

They also served to teach the industrial models later adopted, including mechanization and mass production (later copied and adopted by Ford and General Motors),[25] vertical integration[26] (also later copied and adopted by Ford), aggressive advertising[27] (as much as 10% of all advertising in U.S. periodicals in 1898 was by bicycle makers),[28] lobbying for better roads (which had the side benefit of acting as advertising, and of improving sales by providing more places to ride),[29] all first practised by Pope.[29] In addition, bicycle makers adopted the annual model change[30] [31] (later derided as planned obsolescence, and usually credited to General Motors), which proved very successful.[32]

Furthermore, early bicycles were an example of conspicuous consumption, being adopted by the fashionable elites.[33] In addition, by serving as a platform for accessories, which could ultimately cost more than the bicycle itself, it paved the way for the likes of the Barbie doll.[34]

Moreover, they helped create, or enhance, new kinds of businesses, such as bicycle messengers,[35] travelling seamstresses,[36] riding academies,[37] and racing rinks[38] (Their board tracks were later adapted to early motorcycle and automobile racing.) Also, there were a variety of new inventions, such as spoke tighteners,[39] and specialized lights,[40] socks and shoes,[41] and even cameras (such as the Eastman Company's *Poco*).[42] Probably the best known and most widely used of these inventions, adopted well beyond cycling, is Charles Bennett's Bike Web, which came to be called the "jock strap".[43]

They also presaged a move away from public transit[44] that would explode with the introduction of the automobile.

J. K. Starley's company became the Rover Cycle Company Ltd. in the late 1890s, and then simply the Rover Company when it started making cars. The Morris Motor Company (in Oxford) and Škoda also began in the bicycle business, as did the Wright brothers.[45] Alistair Craig, whose company eventually emerged to become the engine manufacturers Ailsa Craig, also started from manufacturing bicycles, in Glasgow in March 1885.

A man uses a bicycle to carry goods in Ouagadougou, Burkina Faso

In general, U.S. and European cycle manufacturers used to assemble cycles from their own frames and components made by other companies, although very large companies (such as Raleigh) used to make almost every part of a bicycle (including bottom brackets, axles, etc.) In recent years, those bicycle makers have greatly changed their methods of production. Now, almost none of them produce their own frames.

Many newer or smaller companies only design and market their products; the actual production is done by Asian companies. For example, some 60% of the world's bicycles are now being made in China. Despite this shift in production, as nations such as China and India become more wealthy, their own use of bicycles has declined due to the increasing affordability of cars and motorcycles. One of the major reasons for the proliferation of Chinese-made bicycles in foreign markets is the lower cost of labor in China.[46]

One of the profound economic implications of bicycle use is that it liberates the user from oil consumption (Ballantine, 1972). The bicycle is a inexpensive, fast, healthy and environmentally friendly mode of transport (Illich, 1974)

Legal requirements

Reflectors for riding after dark

Early in its development, as with automobiles, there were restrictions on the operation of bicycles. Along with advertising, and to gain free publicity, Albert A. Pope litigated on behalf of cyclists.[47]

The 1968 Vienna Convention on Road Traffic of the United Nations considers a bicycle to be a vehicle, and a person controlling a bicycle (whether actually riding or not) is considered an operator. The traffic codes of many countries reflect these definitions and demand that a bicycle satisfy certain legal requirements before it can be used on public roads. In many jurisdictions, it is an offense to use a bicycle that is not in a roadworthy condition.

In most jurisdictions, bicycles must have functioning front and rear lights when ridden after dark. As some generator or dynamo-driven lamps only operate while moving, rear reflectors are frequently also mandatory. Since a moving bicycle makes little noise, some countries insist that bicycles have a warning bell for use when approaching pedestrians, equestrians, and other cyclists.

Some countries require child and/or adult cyclists to wear Helmets, as this may protect riders from head trauma . Countries which require adult cyclists to wear helmets include Spain, Canada and Australia.

See also

- Cycling – use of bicycles

General

- Bicycle commuting
- Bicycle industry and List of bicycle manufacturing companies
- Bicycle law
- Bicycle lighting
- Bicycle lock
- Bicycle locker
- Bicycle safety
- Bicycle tools
- List of bicycle and human powered vehicle museums
- → List of bicycle parts

Special uses and related vehicle types

- Balance bicycle
- Beach cruiser

- Bicycle trailer
- Boda-boda
- Cycle rickshaw
- Faired bicycle
- Bixi
- Folding bicycle
- Freight bicycle
- Infantry bicycle
- Monowheel
- Quadracycle
- Shaft-driven bicycle
- Tandem bicycle
- Trailer bike
- Tricycle
- Utility cycling
- Unicycle
- Velocipede
- Workbike

Other

- Human-powered transport
- Safety standards
- Transportation technology, timeline of

References

- *All About Bicycling*, Rand McNally.
- Richard Ballantine, *Richard's Bicycle Book*, Pan, 1975.
- Caunter C. F. *The History and Development of Cycles* Science Museum London 1972.
- Daniel Kirshner. *Some nonexplanations of bicycle stability*. American Journal of Physics, 48(1), 1980. The abstract reads "In this paper we attempt to verify a nongyroscopic theory of bicycle stability, and fail".
- David B. Perry, *Bike Cult: the Ultimate Guide to Human-powered Vehicles*, Four Walls Eight Windows, 1995.
- Roni Sarig, *The Everything Bicycle Book*, Adams Media Corporation, 1997
- "Randonneurs USA [48]". *PBP: Paris-Brest-Paris*. March 31 2005.
- US Department of Transportation, Federal Highway Administration. "America's Highways 1776-1976", pp. 42–43. Washington, DC, US Government Printing Office.
- David Gordon Wilson, *Bicycling Science*, MIT press, ISBN 0-262-73154-1
- David V. Herlihy, *Bicycle: The History*, Yale University Press, 2004
- Frank Berto, *The Dancing Chain: History and Development of the Derailleur Bicycle*, San Francisco: Van der Plas Publications, 2005, ISBN 1-892495-41-4.
- *The Data Book: 100 Years of Bicycle Component and Accessory Design*, San Francisco: Van der Plas Publications, 2005, ISBN 1-892495-01-5.
- "Bicycle facts [49]". Retrieved 2006-07-25.

Other authors: Eddie Borysewicz, Greg LeMond, Davis Phinney, Connie Carpenter.

External links

- Pedaling History Museum [50] The world's largest bicycle museum
- P.E.P.A. [51] The official site of Cycling Association of Veteran Athletes of Greece
- Home Made Electric Bicycle [52]
- A range of Traffic Advisory Leaflets [53] produced by the UK Department for Transport covering cycling.
- Menotomy Vintage Bicycles [54] – Databases of antique bicycle photos, features, price guide and research tools. Very large archives.
- The Bicycle - Worlds most efficient form of transportation [55] Discussion of the Bicycle and its advantages over motor vehicles
- Brown, Sheldon [56] (2005). Extensive Online Bicycle Glossary [57]
- Hudson, William (2003). Myths and Milestones in Bicycle Evolution [58]. Retrieved March 30 2005.
- A History of Bicycles and Other Cycles [59] at the Canada Science and Technology Museum
- Jones, David E. H. (1970). The Stability of the Bicycle [60]. Scanned in copy for download for personal use.
- The World Awheel: Early Cycling Books at the Lilly Library [61]
- Bicycle Maintenance [62] :A Wikibooks series

References

[1] DidYouKnow.cd. There are about a billion or more bicycles in the world. (http://www.didyouknow.cd/bicycles.htm) Retrieved 30 July 2006.

[2] Herlihy, David V. (2004). *Bicycle: the history*. Yale University Press. pp. 200–250. ISBN 0-300-10418-9.

[3] " Canada Science and Technology Museum: Baron von Drais' Bicycle (http://www.sciencetech.technomuses.ca/english/collection/cycles2.cfm)". 2006. . Retrieved 2006-12-23.

[4] Norcliffe, Glen. *The Ride to Modernity: The Bicycle in Canada, 1869-1900* (Toronto: University of Toronto Press, 2001), p.50, citing Derek Roberts.

[5] Norcliffe, p.47.

[6] Various (9 December 2006). " Like falling off (http://www.newscientist.com/article/mg19225812.400)". *New Scientist* (2581): 93. . Retrieved 27 January 2009.

[7] Meijaard, Papadopoulos, Ruina, and Schwab, J.P.; Papadopoulos, Jim M.; Ruina, Andy; Schwab, A.L. (2007). "Linearized dynamics equations for the balance and steer of a bicycle: a benchmark and review". *Proc. R. Soc. A.* **463** (2084): 1955–1982. doi: 10.1098/rspa.2007.1857 (http://dx.doi.org/10.1098/rspa.2007.1857).

[8] Wilson, David Gordon; Jim Papadopoulos (2004). *Bicycling Science* (Third ed.). The MIT Press. pp. 270–272. ISBN 0-262-73154-1.

[9] Fajans, Joel (July 2000). " Steering in bicycles and motorcycles (http://socrates.berkeley.edu/~fajans/pub/pdffiles/SteerBikeAJP.PDF)" (PDF). *American Journal of Physics* **68** (7): 654–659. doi: 10.1119/1.19504 (http://dx.doi.org/10.1119/1.19504). . Retrieved 2006-08-04.

[10] Cossalter, Vittore (2006). *Motorcycle Dynamics* (Second ed.). Lulu.com. pp. 241–342. ISBN 978-1-4303-0861-4.

[11] "Bicycle Technology", S.S. Wilson, Scientific American, March 1973

[12] "Johns Hopkins Gazette" (http://www.jhu.edu/~gazette/1999/aug3099/30pedal.html), 30 August 1999

[13] Whitt, Frank R.; David G. Wilson (1982). *Bicycling Science* (Second ed.). Massachusetts Institute of Technology. pp. 277–300. ISBN 0-262-23111-5.

[14] How Much Do Bicycles Pollute? Looking at the Carbon Dioxide Produced by Bicycles (http://www.kenkifer.com/bikepages/advocacy/bike_co2.htm)

[15] History Loudly Tells WhyThe Recumbent Bike Is Popular Today (http://www.recumbent-bikes-truth-for-you.com/history.html)

[16] Lukenbill, Jen (13 February 2008). " Bamboo bikes (http://www.aboutmyplanet.com/environment/bamboo-bikes/)". *AboutMyPlanet.com*. .

[17] Brown, Sheldon. " Fixed Gear Conversions: Braking (http://www.sheldonbrown.com/fixed.html)". . Retrieved 2009-02-11.

[18] Bluejay, Michael. " Safety Accessories (http://bicycleuniverse.info/eqp/accessories.html#safety)". *Bicycle Accessories*. BicycleUniverse.info. . Retrieved 2006-09-13.

[19] " The Essentials of Bike Clothing (http://bicycling.about.com/library/weekly/aa041098.htm)". *About Bicycling*. About.com. . Retrieved 2006-09-13.

[20] " Sheldon Brown: Flat tires (http://sheldonbrown.com/flats.html#patching)". . Retrieved 2008-05-29.

[21] " BikeWebSite: Bicycle Glossary – Patch kit (http://www.bikewebsite.com/bikeop.htm)". . Retrieved 2008-06-20.

[22] Other ISO Technical Committees have published various cycle relevant standards, for example:

- ISO 5775 Bicycle tire and rim designations
- ISO 9633 Cycle chains—Characteristics and test methods

Published cycle standards from CEN TC333 include:

- EN 14764 City and trekking bicycles – Safety requirements and test methods
- EN 14765 Bicycles for young children – Safety requirements and test methods
- EN 14766 Mountain-bicycles – Safety requirements and test methods
- EN 14781 Racing bicycles – Safety requirements and test methods
- EN 14782 Bicycles – Accessories for bicycles – Luggage carriers
- EN 15496 Cycles – Requirements and test methods for cycle locks

Yet to be approved cycle standards from CEN TC333:

- EN 15194 Cycles—Electrically power assisted cycles (EPAC bicycle)
- EN 15532 Cycles—Terminology
- 00333011 Cycles – Bicycles trailers – safety requirements and test methods

[23] http://la.streetsblog.org/2008/10/16/metro-making-room-for-bikes-on-their-trains/
[24] Newnham College Cambridge: The History of the College (http://www.newn.cam.ac.uk/about-newnham/college-history/history/content/history-of-the-college-2)
[25] Norcliffe, Glen. *The Ride to Modernity: The Bicycle in Canada, 1869-1900* (Toronto: University of Toronto Press, 2001), pp.23, 106, & 108. GM's practice of sharing chassis, bodies, and other parts is exactly what the early bicycle manufacturer Pope was doing.
[26] Norcliffe, p.106.
[27] Norcliffe, pp.142–7.
[28] Norcliffe, p.145.
[29] Norcliffe, p.108.
[30] Norcliffe, p.23.
[31] Babaian, Sharon. *The Most Benevolent Machine: A Historical Assessment of Cycles in Canada* (Ottawa: National Museum of Science and Technology, 1998), p.97.
[32] Babaian, p.98.
[33] Norcliffe, pp.8, 12, 14, 23, 147–8, 187–8, 208, & 243–5.
[34] Norcliffe, pp.23, 121, & 123.
[35] Norcliffe, p.212.
[36] Norcliffe, p.214.
[37] Norcliffe, p.131.
[38] Norcliffe, p.30 & 131.
[39] Norcliffe, p.125.
[40] Norcliffe, p.123 & 125.
[41] Norcliffe, p.125 & 126.
[42] Norcliffe, p.238.
[43] Norcliffe, p.128.
[44] Norcliffe, p.214–5.
[45] " The Wrights' bicycle shop (http://www.nasm.si.edu/Wrightbrothers/who/1893/shop.cfm)". 2007. . Retrieved 2007-02-05.
[46] The Economist, 15 February 2003
[47] Norcliffe, Glen. *The Ride to Modernity: The Bicycle in Canada, 1869-1900* (Toronto: University of Toronto Press, 2001), p.108.
[48] http://www.rusa.org/pbp.html
[49] http://www.didyouknow.cd/bicycles.htm
[50] http://www.pedalinghistory.com/
[51] http://www.pepa.gr/en/pepa.html
[52] http://www.greenoptimistic.com/2008/04/16/home-made-electric-bicycle/
[53] http://www.dft.gov.uk/pgr/roads/tpm/tal/cyclefacilities/
[54] http://www.OldRoads.com/
[55] http://gotoes.org/bikestuff/index.htm
[56] http://sheldonbrown.com/
[57] http://sheldonbrown.com/glossary.html
[58] http://www.jimlangley.net/ride/bicyclehistorywh.html
[59] http://www.sciencetech.technomuses.ca/english/collection/cycles.cfm
[60] http://ist-socrates.berkeley.edu/~fajans/Teaching/MoreBikeFiles/JonesBikeBW.pdf
[61] http://www.indiana.edu/~liblilly/awheel/awheel.html
[62] http://en.wikibooks.org/wiki/Bicycles/Maintenance_and_Repair

Freewheel

In mechanical or automotive engineering, a **freewheel** or **overrunning clutch** is a device in a transmission that disengages the driveshaft from the driven shaft when the driven shaft rotates faster than the driveshaft. An overdrive is sometimes mistakenly called a freewheel, but is otherwise unrelated.

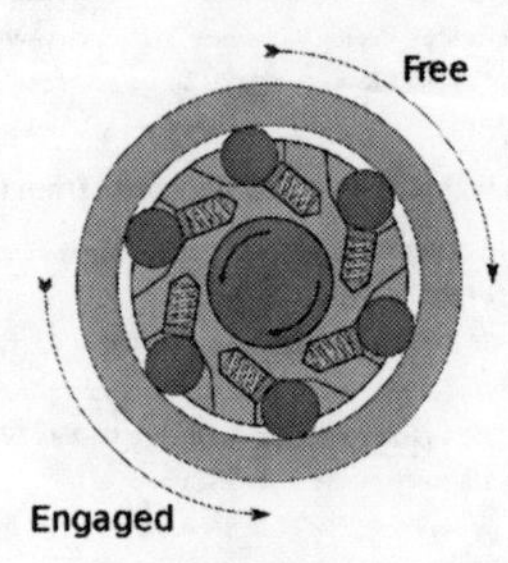

Freewheel mechanism

The condition of a driven shaft spinning faster than its driveshaft exists in most → bicycles when the rider holds his or her feet still, no longer pushing the pedals. Without a freewheel the rear wheel would drive the pedals around.

An analogous condition exists in an automobile with a manual transmission going down hill or any situation where the driver takes his foot off the gas pedal, closing the throttle; the wheels want to drive the engine, possibly at a higher RPM. In a two-stroke engine this is a catastrophic situation: as the engine depends on a fuel/oil mixture for lubrication, a shortage of fuel to the engine would result in a shortage of oil in the cylinders, and the pistons would seize after a very short time causing extensive engine damage. Saab used a freewheel system in their two-stroke models for this reason and maintained it in the Saab 96 V4 and early Saab 99 for better fuel efficiency.

Mechanics

The simplest freewheel device consists of two saw-toothed, spring-loaded discs pressing against each other with the toothed sides together, somewhat like a ratchet. Rotating in one direction, the saw teeth of the drive disc lock with the teeth of the driven disc, making it rotate at the same speed. If the drive disc slows down or stops rotating, the teeth of the driven disc slip over the drive disc teeth and continue rotating, producing a characteristic clicking sound proportionate to the speed difference of the driven gear relative to that of the (slower) driving gear.

A more sophisticated and rugged design has spring-loaded steel rollers inside a driven cylinder. Rotating in one direction, the rollers lock with the cylinder making it rotate in unison. Rotating slower, or in the other direction, the steel rollers just slip inside the cylinder.

Most bicycle freewheels use an internally step-toothed drum with two or more spring-loaded, hardened steel pawls to transmit the load. More pawls help spread the wear and give greater reliability although, unless the device is made to tolerances not normally found in bicycle components, simultaneous engagement of more than two pawls is rarely achieved.

Benefits

By its nature, a freewheel acts as an automatic clutch, making it possible to change gears in a manual gearbox, either up- or downshifting, without depressing the clutch pedal, limiting the use of the manual clutch to starting from standstill or stopping.

A freewheel also produces slightly better fuel economy on carburetted engines (without fuel turn-off on engine brake) and less wear on the manual clutch, but leads to more wear on the brakes as there is no longer any ability to perform engine braking. This makes freewheel transmissions dangerous for use on trucks and automobiles driven in mountainous regions, as prolonged and continuous application of brakes to limit vehicle speed soon leads to brake-system overheating followed shortly by total failure.

Uses

Agricultural equipment

In agricultural equipment an **overrunning clutch** is typically used on hay balers and other equipment with a high inertial load, particularly when used in conjunction with a tractor without a live power take-off (PTO). Without a live PTO, a high inertial load can cause the tractor to continue to move forward even when the foot clutch is depressed, creating an unsafe condition. By disconnecting the load from the PTO under these conditions, the overrunning clutch improves safety. Similarly, many unpowered 'push' cylinder lawnmowers use a freewheel to drive the blades: these are geared or chain-driven to rotate at high speed and the freewheel prevents their momentum being transferred in the reverse direction through the drive when the machine is halted.

Engine starters

A freewheel assembly is also widely used on engine starters as a kind of protective device. Starter motors usually need to spin at 3,000 RPM to get the engine to turn over. When the key is turned to the start position for any amount of time after the engine has already turned over, the starter can not spin fast enough to keep up with the flywheel. Because of the extreme gear ratio between starter gear and flywheel (about 15 or 20:1) it would spin the starter armature at dangerously high speeds, causing an explosion when the centripetal force acting on the copper coils wound in the armature can no longer resist the outward force acting on them. In starters without the freewheel or overrun clutch this would be a major problem because, with the flywheel spinning at about 1,000 RPM at idle, the starter, if engaged with the flywheel, would be forced to spin between 15,000 and 20,000 RPM. Once the engine has turned over and is running, the overrun clutch will release the starter from the flywheel and prevent the gears from re-meshing (as in an accidental turning of the ignition key) while the engine is running. A freewheel clutch is now used in many motorcycles with an electric starter motor. It is used as a replacement for the Bendix drive used on most auto starters because it reduces the electrical needs of the starting system.

Vehicle transmissions

In addition to the automotive uses listed above (i.e. in two-stroke-engine vehicles and early four-stroke Saabs), freewheels were used in some luxury or up-market conventional cars (such as Rovers and Cords) from the 1930s into the 1960s. The freewheel meant that the engine returned to its idle speed on the overrun, thus greatly reducing noise from both the engine and gearbox. The mechanism could usually be locked to provide engine braking if needed. A freewheel was also used in the original Land Rover vehicle from 1948 to 1951. The freewheel controlled drive from the gearbox to the front axle, which disengaged on the overrun. This allowed the vehicle to have a permanent 4 wheel drive system by avoiding 'wind-up' forces in the transmission. This system worked, but produced unpredictable handling, especially in slippery conditions or when towing, and was replaced by a conventional selectable 4WD system.

Bicycles

In the older style of bicycle, where the freewheel mechanism is included in the gear assembly, the system is called a *freewheel*, whereas the newer style, in which the freewheel mechanism is in the hub, is called a *freehub*.

History

The friction freewheel was a part of the Torpedo bicycle gear hub invented by Ernst Sachs in 1903.

A bicycle freewheel was developed and patented by cycle component manufacturer Villiers Engineering of England in 1902.

The freewheel is sometimes known as the "Stieber Clutch" named after its German developer Ortwin Stieber.

See also

- Slipper clutch
- Sprag

Sprocket

A **sprocket** is a profiled wheel with teeth that meshes with a chain, track or other perforated or indented material. It is distinguished from a gear in that sprockets are never meshed together directly, and from a pulley by not usually having a flange at each side.

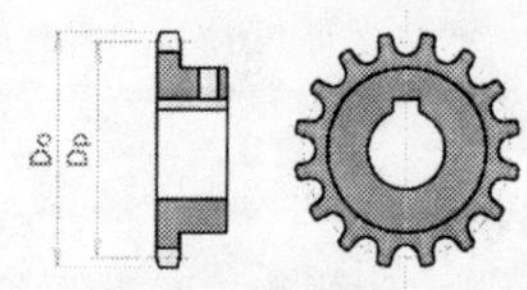

16T type sprocket
Do: Sprocket diameter
Dp: Pitch diameter

Sprockets are used in → bicycles, motorcycles, cars, tanks, and other machinery either to transmit rotary motion between two shafts where gears are unsuitable or to impart linear motion to a track, tape etc.

Cycles

In the case of bicycle chains, it is possible to modify the overall gear ratio of the chain drive by varying the diameter (and therefore, the tooth count) of the sprockets on each side of the chain. This is the basis of Derailleur gears. A 10-speed bicycle, by providing two different-sized driving sprockets and five different-sized driven sprockets, allows up to ten different gear ratios. The resulting lower gear ratios make the bike easier to pedal up hills while the higher gear ratios make the bike faster to pedal on flat roads. In a similar way, manually changing the sprockets on a motorcycle can change the characteristics of acceleration and top speed by modifying the final drive gear ratio.

A sprocket and roller chain

Tracked vehicles

In the case of vehicles with caterpillar tracks the engine-driven toothed-wheel transmitting motion to the tracks is known as the *drive sprocket* and may be positioned at the front or back of the vehicle, or in some cases, both. It can also be a third sprocket, elevated, driving the track.

A tank sprocket

Film and paper

Sprockets are used in the film transport mechanisms of movie projectors and movie cameras. In this case, the sprocket wheels engage film perforations in the film stock.

Sprocket feed was also used for punched tape and is used for paper feed to some computer printers.

External links

- Chain Engagement with Sprockets [1]

See also

- Toothed belt

References

[1] http://chain-guide.com/basics/2-1-2-engagement-with-sprockets.html

Velodrome

A **velodrome** is an arena for → track cycling. Modern velodromes feature steeply banked oval tracks, consisting of two 180-degree circular bends connected by two straights. The straights transition to the circular turn through a moderate easement curve.

The Dunc Gray Velodrome located in the City of Bankstown

Technical aspects

Banking in the turns, called superelevation, allows riders to keep their bikes relatively perpendicular to the surface while riding at speed. When traveling through the turns at racing speed, which may exceed 85 km/h (about 52 mph), the banking attempts to match the natural lean of a bicycle moving through that curve. Therefore, the centripetal acceleration of the combined inertia of bicycle and rider moving in the curved path balances the tangential acceleration pulling them outwards. There is no centrifugal force 'trying' to tilt the bicycle outward, a net normal force is acting on the tires through the riding surface.

Riders are not always traveling at full speed or at a specific radius. Most events have riders all over the track. Team races (like the madison) have some riders at speed and others riding more slowly. In match sprints riders may stop. For these reasons, the banking tends to be 10 to 15 degrees less than physics predicts. Also, the straights are banked 10 to 15 degrees more than physics would predict. These compromises make the track ridable at a range of speeds.

From the straight, the curve of the track increases gradually into the circular turn. This section of decreasing radius is called the easement spiral or transition. It allows bicycles to follow the track around the corner at a constant radial position. Thus riders can concentrate on tactics rather than steering.

Bicycles and track design

Bicycles for velodromes have no brakes. They employ a single fixed rear gear, or cog, that does not freewheel. This helps maximize speed, reduces weight, avoids sudden braking while nevertheless allowing the rider to slow by pushing back against the pedals.

Bicycle racing on an outdoor velodrome.

Modern velodromes are constructed by specialised designers. The Schuermann architects in Germany have built more than 125 tracks worldwide. Most of Schuermann's wooden outdoor tracks are made of wood trusswork with a surface of strips of the rare rain-forest wood Afzelia. Indoor velodromes are built with less expensive pine surfaces. Other designers have been moving away from traditional materials. The 1996 Atlanta Olympics saw the introduction of synthetic surfaces supported by steel frames.

The track is measured along a line 20 cm up from the bottom. Olympic standard velodromes may only measure between 250 m and 400 m, and the length must be such that a whole or half number of laps give a distance of 1 km. Others range from 133 m to 500 m, although 250m is the most popular and the length used in major events. The velodrome at Calshot Spit, Hampshire, UK is only 142 m because it was built to fit inside an aircraft hangar. It has especially steep banking. Forest City Velodrome in London, Ontario, Canada, is the world's shortest at 138 m. It was built to fit a hockey arena. Like Calshot, it has steep banking.

Many old tracks were built around athletics tracks or other grounds and any banking was shallow. The smaller the track, the steeper the banking. A 250 m track banks around 45°, while a 333 m track banks around 32°.

Velodrome tracks can be surfaced with different materials, including wood, synthetics and concrete. Shorter, newer, and Olympic quality tracks tend to be wood or synthetics; longer, older, or inexpensive tracks are concrete, macadam, or even cinder, as in the Little 500.

Track markings

Track markings

All tracks must have standard markings. Between the infield (sometimes referred to as an apron) and the actual track is the blue band (called "côte d'azur") that is typically 10% of the surface. The blue band is not a part of the track. Although it is not illegal to ride there, moving into it to shortcut another rider will result in disqualification. During time trials, pursuits or other timed events, the blue band is obstructed with sponges or other objects. The blue band is a warning to cyclists that they may scrape their pedal along the infield when in a curve. This can easily result in a crash, so this is why it is ill-advised to ride on the blue band.

20 cm above the blue band is the black line. The inner edge of this 5 cm line defines the length of the track. 90 cm above the inside of the track is the outside of the 5 cm wide red sprinter's line. The zone between black and red lines is the optimum route around the track. A rider leading in this zone cannot be passed on the inside; other riders must pass on the longer outside route.

Minimum 250 cm (or half the track width) above the inside of the track is the blue stayers' line. This line serves in races behind motorbikes as a separation line. Stayers below the blue line may not be overtaken on the inside. In Madison races (named after six-day races at Madison Square Garden in New York City, New York and known as the American), the team's relief rider rests above the Stayer's line by riding slowly until his or her teammate comes around the track and throws him or her back into the race.

The finish line is black on white and towards the end of the home straight. Red lines are marked in the middle of each straight as start and finish line for pursuit races. A white 200 m line marks 200 m before the finish.

Track construction

Velodromes may be indoors or outdoors. In the heyday of velodrome racing (1890-1920), indoor tracks were common. When hosting six-day races, they were popular for revelers and urban sophisticates to congregate in the early hours after the bars had closed. Indoor tracks are not affected by weather and are more comfortable for spectators. They ride smoother and last longer. Despite the advantages of indoor tracks, outdoor velodromes are more common, as an outdoor venue does not require a building, making it more affordable, especially when new. Today, although many classic indoor tracks have been torn out of buildings and replaced by venues for more popular sports, velodromes are still sometimes built into indoor venues, particularly where track racing can generate enough to cover the expense of dedicating a building to it.

A velodrome will usually be among facilities constructed for events such as the Olympics or Commonwealth Games.

Race Formats

There are a variety of formats in velodrome races. A typical event will consist of several races of varying distances and structures. The most straight forward is the "**scratch race**," where riders compete over a specified distance and the order of finish determines the winners.

Points races assign value to specific laps throughout a race and riders position in relation to the field, generally the leading rider and occassionally the second place rider will be awarded points. The structure and timing of points races varies greatly, but the winner is determined by the accumulation of points and not necessarily the rider crossing the line first at the end of the race.

Elimination race, also known as "devil take the hindmost" removes the last place rider from each lap until only three to five riders remain. The final standings are then determined by a sprint over the last two laps.

Madison races team up pairs of riders in a tag-team format. Riders "sling" their teammate forward to facilitate alternating sprints that keep the pace very high during typically long races (30km or more compared to 3-10km for most other races). The name is taken from Madison Square Garden where the format was popular in the early 1900's.

Keirin races involve pacing 6-9 riders with a motorcycle until the last lap and a half when a sprint for the finish determines the winner.

Omnium competition assigns a point value to final standings of each race and riders accumulate points over the course of an event or series of events. This is not a specific race, but a competition that ties races and events together.

[1]

See also

- List of velodromes
- → Track cycling

External links

- List of velodromes [2]

References

[1] http://www.thevelodrome.com
[2] http://www.bikecult.com/bikecultbook/sports_velodromes.html

Track bicycle

A **track bicycle** or **track bike** is a bicycle optimized for racing at a → velodrome or outdoor track. Unlike road bicycles, the track bike is a → fixed-gear bicycle and so has a single gear and neither → freewheel nor brakes. Tires are narrow and inflated to high pressure to reduce rolling resistance. Tubular tires are often used.

A track bicycle

Frame design

A track frame is specific to its use. Rigidity is more important than lightness. Frames for sprinting are as rigid as possible, while those for general racing as aerodynamic as possible.

Rules

The governing body, the International Cycling Union (UCI), sets limits on design and dimensions as well as the shape and diameter of the tubes used to construct the frame.

Geometry

A track bicycle differs from one used on the road by having:

- higher bottom bracket so the pedals do not touch a steeply banked track
- steeper seat tube for a more powerful aerodynamic position,
- steeper head tube for more responsive steering,
- less fork rake.

Typical track frames use 120mm spacing for the rear hub. The dropouts or *track ends* face rearwards to facilitate chain tension adjustment with very tight clearances in front of the rear tire that would prevent wheel removal with forward facing dropouts.

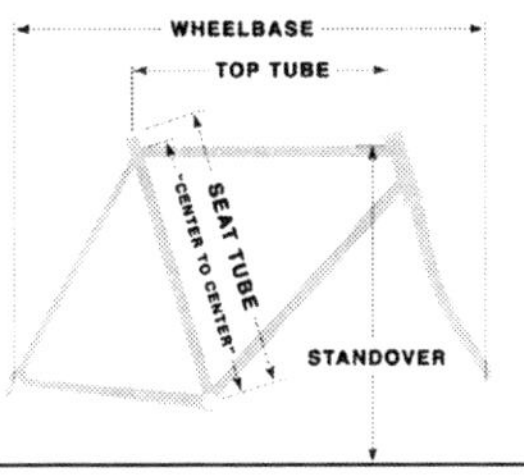

Bicycle frame measurements

Material

Frames can be made of steel, aluminium, carbon fiber, or titanium. Carbon fiber is most common at the professional level.

Gears

Track bicycles have only one gear so its size is important. A lower gear allows quicker acceleration or 'jump.' But a bigger gear makes sustained speed easier, important in pursuit, time trial and bunched races such as points or scratch events. Without a good jump, the rider risks opponents accelerating away; without good sustained speed, he will be unable to keep up with a fast race. Track cyclists practice fast pedalling (cadence) as a compromise.

Long-distance attempts such as the hour record use high gear combinations such as 52x12 or 55x14. Ondřej Sosenka used 54x13 with 190mm cranks to set the 2005 record.

Chain

There are two common widths of single speed and fixed gear bicycle chains: 1/8 inch and 3/32 inch. The chainring, sprocket and chain should all be the same width. Although an 1/8-inch chain will work on a 3/32-inch chainring or sprocket, it is not ideal. A 3/32-inch chain will not work on a 1/8-inch chainring or sprocket. Because they do not need to shift from sprocket to sprocket, track chains use a full bushing to allow little flex and to be stronger. All bicycles with derailleur gears use bushingless chains which flex, making gear changing possible.

Bicycle messenger

Bicycle messengers (also known as bike or cycle couriers) are people who work for courier companies (also known as messenger companies) carrying and delivering items by → bicycle. Bicycle messengers are most often found in the central business districts of metropolitan areas. Courier companies use bike messengers because bicycle travel is less subject to unexpected holdups in city traffic jams, thereby offering a predictable delivery time.

Bicycle courier, London, UK

History

Almost immediately after the development of the pedal-driven velocipede in the 1860s, people began to use the bicycle for delivery purposes. David Herlihy's 2004 book on the early history of the bicycle contains several references to bicycle messengers working during the late 19th century, including a description of couriers employed by the Paris stock exchange in the 1870s.[1] During the bicycle boom of the 1890s in the United States, Western Union employed a number of bicycle messengers in New York City and other large population centers.

Bicycle messenger boys, Salt Lake City, 1912

The earliest recorded post-war American bicycle courier company was founded by Carl Sparks, in San Francisco 1945. According to the San Francisco Bicycle Messenger Association, "Sparkie's went on to become Aero, which was bought out in 1998 [and] later absorbed into CitySprint."[2] By the late 1970s, there were well-established companies offering bicycle messenger services in many major cities in the U.S.

In Europe, the bicycle had fallen out of favour as a means of delivery in the third quarter of the 20th century. It was not until 1983 that bicycle messengers made their reappearance in Europe. London's "On Yer Bike" and "Pedal-Pushers" were pioneers of pedal over petrol, and the rest of the city's courier companies followed suit. By the late 1980s, cycle couriers were a common sight in central London and a British manufacturer named a range of mountain-bikes for them, the Muddy Fox 'Courier'.[3] Entrepreneurs in continental Europe, some inspired by seeing couriers in the U.S. or in London, began to offer bicycle courier services in the late 1980s, and by 1993 there were sufficiently large numbers of bicycle couriers in Northern Europe and North America that over 400 attended the inaugural Cycle Messenger Championships in Berlin, Germany.[4] Bicycle messengers have not become common in southern Europe, the heartland of world competitive cycling. There are very few bicycle couriers in Portugal, France, Spain, or Italy. Outside Europe and North America, there are now large bicycle messenger services in Japan, and also in New Zealand and Australia.

Demand for courier services

Messengers carry a huge variety of items, from things that could not be sent by digital means (corporate gifts, original artwork, clothes for photo-shoots, original signed documents) to mundane items that could easily be emailed, albeit without the air of importance attached to an express courier delivery. Messengers deliver digital content on optical media or hard disks because, despite high speed broadband connections, companies find it easier to send a disc than to work out how to transmit larger amounts of data than an email account can handle.[5] [6] Legal documents, various financial instruments and sensitive information are routinely sent by courier, reflecting a distrust of digital cryptography.[7]

Commentators have claimed that technological innovation will significantly reduce the demand for same-day parcel delivery, [8] [9] predicting that the fax machine, and then the internet, would render the messenger business obsolete. There is still a demand for fast courier services[10] but there is certainly some truth in the predictions. Reliable data specifically relating to bicycle messenger occupational statistics is hard to find: the U.S. Department of Labor statistics do not track bicycle messengers specifically, and do not include "independent contractors" in statistics referenced for this industry occupation,[11] but reports indicate the business is shrinking.

The gradual acceptance of electronic filing by U.S. courts has had a negative effect on the market. In San Francisco bike messengers report a smaller work force coupled with decreased earnings. [12] In New York alone, the number of messengers has been estimated to have dropped by a thousand over the last decade.[13]

Working conditions

The conditions of employment of bicycle messengers vary from country to country, city to city and even company to company. Contracts governing the relationship between individual courier and company are subject as much to customary practice, as local ordinance. In some places messengers are independent contractors paid on commission and do not receive benefits such as health insurance. In other places they will be regular employees of the courier company enjoying all the benefits thereof.

The employment status of the bicycle messengers of one of the UK's biggest same-day courier services, CitySprint, was challenged by the GMB trade union in December 2007. The challenge arose after the firm terminated the contract of one of its riders. The GMB sought to establish that more than 1500 CitySprint operatives currently classified as self-employed sub-contractors should be re-classified as employees.[14] Recent US legislation[15] has sought to force companies to reclassify workers that they declare as independent contractors, as employees, with all the benefits thereof. Companies claim costs will rise as a direct result.[16]

The job is poorly paid relative to the risk and effort required. In 2002, a Harvard Medical School study of injury rates amongst Boston bicycle messengers determined that the rate of injury requiring time off work amongst the sample group was more than 13 times the U.S. average, and more than three times higher than the next highest group, workers in the meat-packing industry.[17] At least one bicycle messenger is killed while working every year in the U.S.[18] [19] [20] Eight bicycle messengers are known to have been killed while working in London between 1989 and 2003.[21] Because payment is made at piece rates, and cycle couriers often fail to file accurate tax returns, it is hard to get reliable figures for messenger income. A study published in 2006 stated that the average daily wage of London bicycle messengers was £65 a day, and that of bicycle messenger in Cardiff was £45.[22] The UK legal minimum wage at that time was £5.52 an hour.[23]

Licensing

In most cities there are no legal requirements beyond those applying to all cyclists. Some jurisdictions, though, require the licensing of courier bicycles. In Calgary, Alberta for example, metal license plates must be fixed to bicycles used for courier work.[24] Vancouver, British Columbia additionally requires bicycle messengers to complete a test in order to obtain a license, and to carry an identification card.[25]

Equipment

The essential equipment for a bicycle messenger is a bicycle. Messengers can be found using many different types of machine, including road bikes, hybrids, mountain bikes, BMX bicycles and → track bicycles. Fixed-gear bicycles, contrary to popular belief, are ridden by a minority of bicycle messengers.[26]

London bicycle courier, 2007. Note lock carried around waist.

The majority of messengers use a bag to carry deliveries and personal effects. Bags with a single strap that wraps diagonally across the chest (popularly known as messenger bag) are popular because they can be swung around the messenger's body to allow access without removing the bag. Clasps which can be adjusted with one hand (ideal for riding), clips, pockets and webbing loops on the strap for holding a cell phone or two-way radio and other equipment also feature on purpose-built messenger bags. Bags generally have large capacities (up to 50 liters or 3,000 cubic inches). Baskets and racks mounted on the bike are also used, and at least one messenger service (in New York City) equips its riders with specialized three-wheel cycles (sometimes known as cargo-trikes), with a large trunk in the rear.

Because bicycle thefts are prevalent in many cities[27] , a lock to secure the bike during deliveries is essential. Simple chain and padlocks are often used, with the locked chain worn around the waist like a belt while riding. U-locks are also popular.

Messengers typically carry basic tools, weather-proof clothing and a street map.[28]

Communications

Messengers communicate and are dispatched to assignments via hand-held communication devices including two-way radios, cell phones, and personal digital assistants. Many of the larger messenger services equip their riders with GPS tracking devices[29] , for ease of location.

Messenger culture and influence

Media

Messengers have been used in fiction media as symbols of urban living, and have been the subject of novels,[30] , memoirs[31] feature films,[32] television series,[33] comic-books,[34] , and sociological studies.[35] Mexican artist José Guadalupe Posada created a popular icon of a marijuana-smoking bicycle courier everyman in his 19th century engravings.

The 1986 film Quicksilver featuring a bicycle messenger, is an early expression of the mythology of the messenger as daredevil stunt rider[36] . Star Kevin Bacon rides a racing bike, a fixed gear bike and a trick bike with 1:1 gear ratio and zero-rake forks, to perform the stunts in the film.

News media have made portrayals of messengers ranging from innocuous urban libertines to reckless, cliquish nihilists. The latter portrayal is often sparked by local incidents involving bike messengers in collisions with other road-users or run-ins with authority figures.[37] These incidents also occasionally lead to proposals for, and dispute over, new ordinances and regulations on messengers and messengering.[38] [39]

Fashion

The influence of bicycle messengers can be seen in urban fashion, most notably the popularity of single-strap messenger bags, which are a common accessory among people who do not ride a bicycle regularly. The rise in popularity of fixed-gear bicycles in the mid-2000s, complete with affectations such as spoke cards (gathered from "alleycats" typically), is attributed to bicycle messengers.[40]

Events

Since 1993, Cycle Messenger Championships have taken place at national, continental and world levels.[41] In addition to these international races, bicycle messengers organise events of many kinds. These range from weekend events featuring multiple competitions,[42] to roller races[43] held in bars, to alleycats and social rides. These events are held as much for fun and messenger networking as for competition. Bicycle messengers also take part in formal cycle competitions at all levels, and in all disciplines. Nelson Vails, silver medallist on the velodrome in the 1984 Olympics, worked as a bicycle messenger in New York City in the early 1980s. Ivonne Kraft, who competed in the 2004 Olympic cross country mountain bike race, is a multiple former Cycle Messenger World Champion, and worked as a bicycle messenger in Germany for a number of years.

The Swiss cycle messenger championships under way in 2004.

See also

- Alleycat races
- Bicycle culture
- Courier

- Freight bicycle
- Newsboys Strike of 1899
- Package delivery

External links

- The International Federation of Bicycle Messenger Associations [44]
- IWW couriers union [45]
- Bike messenger comic [46]

References

[1] Herlihy, David V (2004). *Bicycle: the History* (http://yalepress.yale.edu/yupbooks/book.asp?isbn=9780300104189). Yale University Press. pp. 177. ISBN 0-300-10418-9. .

[2] " A History of Bike Messengering in San Francisco (http://www.ahalenia.com/sfbma/history.html)". San Francisco Bicycle Messenger Association. September 1996. . Retrieved 2007-10-01.

[3] " Value Judgement (http://www.ileach.co.uk/post/archive/archivepost12.html)". The Washing Machine Post. 1998-04-11. . Retrieved 2007-10-01. "When I entered the grand world of cycling in a proper way instead of playing on a terrible ten speed racer, i bought a Muddy Fox Courier for £300 [sic]"

[4] " A Short history of the Cycle Messenger World Championships (http://www.messengers.org/ifbma/history.html)". International Federation of Bicycle Messengers and Associations. October 1998. . Retrieved 2007-10-01. "500 messengers showed up, raced, drank beer, smoked and hung out. The Cycle Messenger World Championships had been born."

[5] " Piracy And Securing The Digital Media Supply Chain (http://publications.mediapost.com/index.cfm?fuseaction=Articles.showArticleHomePage&art_aid=73253)". Video Insider. 2007-12-31. . Retrieved 2008-01-05.

[6] " Sneakernet Redux: Walk Your Data (http://www.wired.com/news/culture/0,1284,54739,00.html)". Wired Magazine. 2002-08-26. . Retrieved 2007-09-19.

[7] " Authenticity and Integrity in the Digital Environment (http://www.clir.org/pubs/reports/pub92/lynch.html)". Council on Library and Information Resources. 2000-05-01. .

[8] Seaton, Matt (2007-07-05). " Two Wheels column (http://www.guardian.co.uk/g2/story/0,,2118772,00.html)". Guardian Media. . Retrieved 2007-09-17.

[9] Tomasson, Robert (1990-03-19). " Fax Displacing Manhattan Bike Couriers (http://query.nytimes.com/gst/fullpage.html?res=9D0CE5D9123CF93AA25750C0A967958260)". New York Times. . Retrieved 2007-09-17.

[10] " Soft Pedal (http://twocitiestwowheels.blogspot.com/2006/07/economist-and-bike-messengers.html)". The Economist. 2006-06-29. .

[11] " US Department of Labor (http://www.bls.gov/oes/current/oes435021.htm)". Bureau of Labor Statistics. . Retrieved 2009-01-09.

[12] Pender, Kathleen. " Bicycle messengers are pedaling uphill against the Internet (http://www.sfgate.com/cgi-bin/article.cgi?f=/c/a/2007/07/17/BUGOSR1ET51.DTL)". San Francisco Chronicle. . Retrieved 2009-01-09.

[13] " Internet Endangers Big-City Tradition: The Bike Messenger (http://www.wired.com/culture/lifestyle/news/2008/07/bikemessengers)". Wired Magazine. 2008-07-25. . Retrieved 2008-07-25.

[14] Nick Mathiason (2007-12-23). " Christmas sacking starts fight for rights of dispatched courier (http://www.guardian.co.uk/business/2007/dec/23/2)". London, UK: Guardian Media. . Retrieved 2007-12-24.

[15] " Obama introduces pro-labor legislstion today (http://my.barackobama.com/page/community/post/unionsforchange/CczH)". BarackObama.com. 2007-09-12. . Retrieved 2007-10-20.

[16] Ron Da Parma (2007-12-27). " IRS says FedEx may owe $319 million (http://www.pittsburghlive.com/x/pittsburghtrib/business/s_544387.html)". Pittsburgh Tribune-Review. . Retrieved 2008-01-03.

[17] Dennerlein, Meeker (2002-02-06). *Occupational Injuries Among Boston Bicycle Messengers* (http://www.hsph.harvard.edu/ergonomics/bike/index.html). Harvard School of Public Health. . Retrieved 2007-09-17.

[18] Greg Ehrendreich (2007-08-15). " Tragedy strikes the CCU (http://www.iww.org/en/node/3578)". www.iww.org. . Retrieved 2009-01-27. ""a bicycle messenger in the Chicago Couriers Union (IU 540), was killed on the job yesterday.""

[19] Emily Vasquez (2006-08-11). " Manhattan: Bike Messenger Killed (http://query.nytimes.com/gst/fullpage.html?res=9B03E7DB163EF932A2575BC0A9609C8B63&sec=&spon=&scp=2&sq=New York: Manhattan: Bike Messenger Killed&st=cse)". New York Times. . Retrieved 2007-09-23.

[20] " Thomas McBride (http://www.ahalenia.com/memorial/mcbride.html)". Messenger Memorial. 1999. . Retrieved 2007-09-23. "d. 26.April.1999 murdered by Carnell Fitzpatrick, driver of a Chevy Tahoe."

[21] 'Buffalo' Bill Chidley (2005-10-13). " Dedication (http://www.movingtargetzine.com/memorial)". Moving Target. . Retrieved 2007-09-23. "London Bicycle Messengers killed while working: Joe Cooper, Calvin Simpson, Paul Ellis, Edward Newstead, Mark Francis, Reidar "Danny" Farr, Sebastian Lukomski"

[22] Fincham, Ben (2007). "'Generally speaking people are in it for the cycling and the beer': Bicycle couriers, subculture and enjoyment". *The Sociological Review* **55** ((2)): 192. doi: 10.1111/j.1467-954X.2007.00701.x (http://dx.doi.org/10.1111/j.1467-954X.2007.00701.x).

[23] " National Minimum Wage (http://www.dti.gov.uk/employment/pay/national-minimum-wage/index.html)". UK Department for Business, Enterprise & Regulatory Reform. 2005-10-13. . Retrieved 2007-12-30.

[24] " Calgary City Business License information FAQ (http://www.calgary.ca/portal/server.pt/gateway/PTARGS_0_0_104_0_0_35/http;/content.calgary.ca/CCA/City+Hall/Business+Units/Development+and+Building+Approvals+and+Land+Use+Planning+and+

Policy/Business+Licences/Business+Licence+FAQs.htm#Question4)". City of Calgary. . Retrieved 2009-01-18.
[25] " City of Vancouver Community Services/Licenses and Inspections (http://vancouver.ca/commsvcs/licandinsp/licences/bikecouriers.htm)". City of Vancouver. . Retrieved 2009-01-22.
[26] " Don't kill the bike messenger (http://www.taipeitimes.com/News/sport/archives/2005/06/11/2003258924)". NY Times News Service. 2005-06-11. .
[27] " Bicycles stolen every 71 seconds (http://news.bbc.co.uk/1/hi/england/6596559.stm)". BBC News. 2007-04-27. . Retrieved 2007-09-24. "The hotspots for [bicycle] thefts are central London"
[28] Chidley and others, 'Buffalo' Bill. " How to be a messenger and not get stitched up, nicked or run over (http://www.londonmessengers.org/FAQ)". London Bicycle Messenger Association. . Retrieved 2007-09-17.
[29] Chidley, 'Buffalo' Bill. " The End of Bicycle Messengering As We Know It? (http://www.movingtargetzine.com/article/the-beginning-of-the-end-of-bicycle-messengering-as-we-know-it)". Moving Target. . Retrieved 2007-09-17.
[30] Gibson, William (1994-10-06). *Virtual Light*. London: Viking Press. ISBN 978-0140157727.
[31] Reilly, Rebecca (2000-08). *Nerves of Steel : Bike Messengers in the United States*. New York: Spoke and Word. ISBN 978-0970342607.
[32] Quicksilver, 1986 film about a stock broker who becomes a NYC messenger
[33] " Streetwise, a UK children's TV series about a cycle courier company (http://www.imdb.com/title/tt0168380/)". . Retrieved 2007-09-18.
[34] Jones, Jay (w), Jones, Jay (p). *Messenger 29* **1** (1) (September 1989), September Press
[35] Fincham, Ben (2007). "'Generally speaking people are in it for the cycling and the beer': Bicycle couriers, subculture and enjoyment". *The Sociological Review* **55** ((2)): 189–202. doi: 10.1111/j.1467-954X.2007.00701.x (http://dx.doi.org/10.1111/j.1467-954X.2007.00701.x).
[36] Online article in the Pittsburg Post-Gazette (http://www.post-gazette.com/pg/07252/815585-109.stm.)
[37] Mahtesian, Charles. " Messenger mayhem (http://www.governing.com/archive/1999/jun/bike.txt)". Governing Magazine. . Retrieved 2009-01-13.
[38] McNulty, Timothy (1999-09-08). " Messengers don't like Council's signals (http://www.post-gazette.com/regionstate/19990908messenger1.asp)". Pittsburgh Post Gazzette. . Retrieved 2007-09-19.
[39] " An Ordinance Concerning The Registration Of Commercial Bicycle Services And The Licensing Of Commercial Messengers (http://www.mass.gov/legis/laws/seslaw98/sl980302.htm)". Government of Massachusetts. . Retrieved 2009-01-18.
[40] Ryan, Singel. " Fixed-Gear Bikes an Urban Fixture (http://www.wired.com/culture/lifestyle/news/2005/04/67149)". Wired Magazine. . Retrieved 2009-01-18.
[41] IFBMA. " CYCLE MESSENGER WORLD CHAMPIONSHIPS Event Listings (http://www.messengers.org/events/)". IFBMA. . Retrieved 2009-01-27.
[42] " Berlin Annual International Cycle Courier Team Trophy (http://www.courier-cup.de/cup/)". . Retrieved 2007-09-17.
[43] " Rollapaluza, the London messenger roller race (http://www.londonmessengers.org/rollapaluza-vii)". . Retrieved 2007-09-17.
[44] http://www.messengers.org
[45] http://www.chicagocouriersunion.org/
[46] http://www.hardabud.com/

Bicycle brake systems

Bicycle brake systems are used to slow down or brake a → bicycle. There have been various types through history, and several are still in use today.

Animation of a single pivot side-pull calliper brake for the rear wheel of a steel framed road bike.

History

Early bicycles such as the high-wheeled penny-farthing bikes were fitted with spoon brakes. As they were fixed gear bicycles, a rider also could reduce speed by reversing the force on the pedals. Unsurprisingly there were many accidents, some fatal, which limited the appeal of cycling mostly to young and adventurous men.

The 1870s saw the development of the "safety bicycle" which roughly resembles bicycles today, with two wheels of equal size, initially with solid rubber tires. These were generally equipped with a front spoon brake and no rear brake, like the penny-farthings fixed gears, allowing control of speed by control of pedalling. Spoon brakes were not very powerful and potentially dangerous in wet weather. A calliper brake was patented by Browett and Harrison in 1887.[1]

With the invention of pneumatic tires in the 1890s came the **rim brake**, the type of brake most commonly used on bicycles today. However, in America throughout most of the 20th century, the most common type of brake was the **coaster brake**, engaged by pressing backwards on the pedals. The rim brake began to supersede the coaster brake in the 1970s.

Brake types

Spoon brakes

Improvised spoon brake on a Chinese cargo tricycle

The **spoon brake** was probably the first type of bicycle brake and precedes the pneumatic tire. Spoon brakes were used on penny farthings with solid rubber tires in the 1800s and continued to be used after the introduction of the pneumatic tired safety bicycle. The spoon brake consists of a pad (often leather) which is pressed onto the top of the front tire. These were almost always rod-operated by a right-hand lever. In developing countries, a foot-operated form of the spoon brake sometimes is retrofitted to old rod brake roadsters. It consists of a spring-loaded flap attached to the back of the fork crown. This is depressed against the front tire by the rider's foot.

Perhaps more so than any other form of bicycle brake, the spoon brake is sensitive to road conditions and increases tire wear dramatically.

Though made obsolete by the introduction of the coaster brake and rod brake, spoon brakes continued to be used in the West supplementally on adult bicycles until the 1930s, and on children's bicycles until the 1950s. In the developing world, they were manufactured until much more recently.

Rim brakes

Rim brakes are so called because braking force is applied by friction pads to the rim of the rotating wheel, thus slowing it and the bicycle. Brake pads can be made of leather or rubber and are mounted in metal "shoes". Rim brakes are typically activated by the rider squeezing a lever mounted on the handlebar.

Advantages and disadvantages

Rim brakes are cheap, light, mechanically simple, easy to maintain, and powerful. However, they perform poorly when the rims are wet. This problem is less serious with rims made of aluminum, found on more expensive bikes, than on those with steel or chromed rims. Rim brakes are also prone to clogging with mud, particularly when mountain biking.

Rim brakes require regular maintenance. Brake pads can wear down quickly, and have to be replaced. Over longer time and use, rims become worn. Rims should be checked for wear periodically as they can fail catastrophically if the rim sidewalls become too worn. Depending on the brake pads and rim, this can happen after a few thousand miles if heavily used in wet and muddy conditions. Bowden cables can become sticky if not regularly lubricated or if water gets into the housing, causing corrosion, although modern lined and stainless steel cables are less prone to these problems. The cables also can wear through repeated use over a long time, however they are more likely to get damaged through getting kinked or the open end becoming unraveled. If the inner cables are not replaced when they fray, they can suddenly break when brakes are applied strongly, causing brakes to be lost when they are most needed. Rim brakes also require that the rim be relatively straight; if the rim has a pronounced wobble, then either the brake pads rub against it when the brakes are released, or apply insufficient and uneven pressure to the rim when certain brakes e.g. dual pivot, are applied.

Rim brakes also heat the rim because the brake functions by converting kinetic energy into thermal energy. In normal use and with lightweight bicycles this is not a problem, as the brakes are only applied with a limited force and for a short time, so the heat quickly dissipates to the surrounding air. However, on heavily-laden touring bikes and tandems in mountainous regions, the heat build-up over a long descent can increase tire pressure so much that the tire blows off the rim. If this happens on the front wheel, a serious accident is almost inevitable. The problem is worse when descending cautiously at slow speeds because the brakes are "always on" and the cooling airflow over the rim is insufficient. The risk can be reduced by not over-inflating tires and adopting an aggressive riding style, only braking for the corners. For this reason, disc brakes are often fitted instead of rim brakes in this situation. Another common solution is to use a drum brake, activated constantly for long descents, in addition to the rim brake.

There are many designs of brake pads (brake blocks). Most consist of a replaceable rubber pad held in a metal channel (brake shoe), with a post or bolt protruding from the back to allow attachment to the brake. Some are made as one piece with the attachment directly molded in the pad for lower production costs. The rubber can be softer for more braking force with less lever effort, or harder for longer life. The rubber can also contain abrasives for better braking, at the expense of rim wear. Compounds vie for better wet braking efficiency. Typically pads are relatively short, but longer varieties are also manufactured to provide more surface area for braking; these often must be curved to match the rim. A larger pad does not give more friction but wears more slowly, so a new pad can be made thinner, simplifying wheel removal with linear-pull brakes in particular. In general, a brake can be fitted with any of these many varieties of pads, as long as the pad mounting method is compatible. Carbon rims, as on some disc wheels, generally have to use non-abrasive cork pads.

The following are among the many sub-types of rim brakes:[2]

Rod-actuated brakes

The **rod-actuated brake**, or simply **rod brake**, uses a series of rods and pivots (rather than Bowden cables) to transmit force applied to a hand lever to pull friction pads upwards against the inner surface (facing the hub) of the wheel rim. They were often called **stirrup brakes** due to their shape. Rod brakes are used with a rim profile known as the Westwood rim, which has a slightly concave area on the braking surface and lacks the flat outer surface required by brakes that apply the pads on opposite sides of the rim.

Front rod brake system

The back linkage mechanism is complicated by the need to allow rotation where the fork attaches to the frame. Although heavy and complex, the linkages are reliable and durable and can be repaired or adjusted with simple hand tools. The design is still in use, typically on roadsters, particularly in East, South Asia and Africa (boda-boda).

The Calliper Brake Design

The calliper brake is a class of cable-actuated brake in which the brake mounts to a single point above the wheel, theoretically allowing the arms to auto-centre on the rim. Arms extend around the tire and end in brake shoes that press against the rim. While some designs incorporate dual pivot points, the arms pivot on a sub-frame — the entire assembly still mounts to a single point.

Calliper brakes are generally considered unsuited to bicycle suspensions, though implementations do exist. They also tend to become less effective as tires get wider. Thus calliper brakes are rarely found on modern mountain bikes. But they are almost ubiquitous on road bikes, particularly the dual-pivot side-pull calliper brake.

Side-pull calliper brakes

Single-pivot side-pull calliper brakes consist of two curved arms that cross at a pivot above the wheel and hold the brake pads on opposite sides of the rim. These arms have extensions on one side, one attached to the cable, the other to the cable housing. When the brake lever is squeezed, the arms move together and the brake pads squeeze the rim.

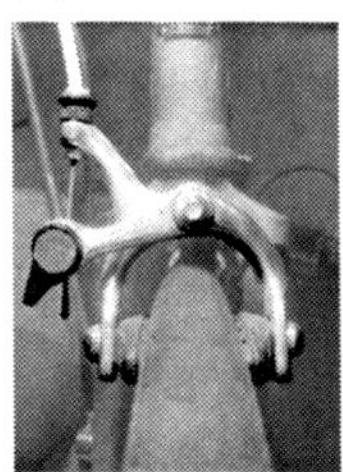
Single pivot side-pull calliper brake.

These brakes are simple and effective for relatively narrow tires but have significant flex and resulting poor performance if made big enough to fit wide tires. Low-quality varieties also tend to rotate to one side during actuation and to stay there, so that one brake pad continually rubs the rim. These brakes are now used on inexpensive bikes; before the introduction of dual-pivot calliper brakes they were used on all types of road bikes.

Dual-pivot calliper brake.

Dual-pivot side-pull calliper brakes are used on most modern racing bicycles. One arm pivots at the centre, like a side-pull; and the other pivots at the side, like a centre-pull. The cable housing attaches like that of a side-pull brake.

The centring of side-pull brakes was improved with the mass-market adoption of dual-pivot side-pulls (an old design re-discovered by Shimano in the early 1990s). These brakes offer a higher mechanical advantage, and resulting better braking. Dual-pivot brakes are slightly heavier than conventional side-pull callipers and cannot accurately track an out-of-true rim.

Centre-pull calliper brakes

Centre-pull calliper brakes have symmetrical arms and as such centre more effectively. The cable housing attaches to a fixed cable stop attached to the frame, and the inner cable bolts to a sliding piece or a small pulley, over which runs a *straddle cable* connecting the two brake arms. Tension on the cable is evenly distributed to the two arms, preventing the brake from taking a "set" to one side or the other.

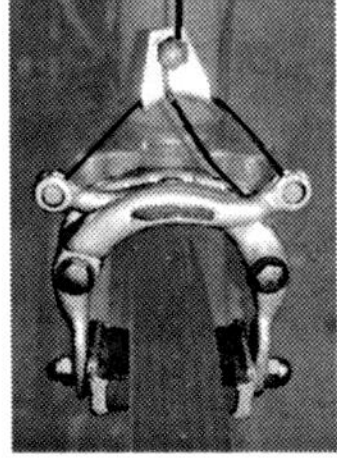
Centre-pull calliper brake.

These brakes were reasonably priced, and in the past filled the price niche between the cheaper and the more expensive models of side-pull brakes.

U-brakes

U-brakes (also known by the trademarked term **"990-style"**) are essentially the same design as the centre-pull calliper brake. The difference is that the two arm pivots attach directly to the frame or fork while those of the centre-pull calliper brake attach to an integral bridge frame that mounts to the frame or fork by a single bolt. Like roller cam brakes, this is a first-class cantilever design with pivots located above the rim. Thus U-brakes are often interchangeable with, and have the same maintenance issues as, roller cam brakes.

U-brakes were used on mountain bikes through the early 1990s, particularly in the then-popular under-the-chainstays rear location;[3]

They are the current standard on Freestyle BMX frames and forks. The U-brake's main advantage over cantilever and linear-pull brakes in this application is that sideways protrusion of the brake and cable system is minimal, and the exposed parts are smooth. This is especially valuable on freestyle BMX bikes where any protruding parts are liable to damage and may interfere with the rider's body.

The Cantilever Brake Design

The cantilever brake is a class of brake in which each arm is attached to a separate pivot point on one side of the seat stay or fork. Thus all cantilever brakes are dual-pivot. Both first and second-class lever designs exist; second-class is by far the most common. In the second class lever design, the arm pivots below the rim. The brake shoe is mounted above the pivot and is pressed against the rim as the two arms are drawn together. In the first class lever design, the arm pivots above the rim. The brake shoe is mounted below the pivot and is pressed against the rim as the two arms are forced apart.

Cantilever brakes are preferred for bicycles that use wide tires, such as those on mountain bikes. (Standard calliper brakes are problematic in these applications since the long distance from the pivot to the pad allows the arms to flex, reducing braking effectiveness.) Because the arms move only in their designed arcs, the brake shoe must be adjustable in several planes. Thus cantilever brake shoes are notoriously difficult to adjust.

There are several brake types based on the cantilever brake design: cantilever brakes and direct-pull brakes - both second class lever designs - and roller cam brakes and U-brakes - both first class lever designs.

Traditional Cantilever brakes

The **traditional cantilever brake**, or commonly **cantilever brake**, pre-dates the direct-pull brake. It is a centre-pull cantilever design with an outwardly-angled arm protruding on each side, a cable stop on the frame or fork to terminate the cable housing, and a straddle cable between the arms similar to centre-pull calliper brakes. The cable from the brake lever pulls upwards on the straddle cable, causing the brake arms to rotate up and inward thus squeezing the rim between the brake pads.

Low profile 'traditional' cantilever brake.

Traditional cantilever brakes are difficult to adapt to bicycle suspensions and protrude somewhat from the frame. Accordingly they are usually found only on bicycles without suspension.

V-brakes

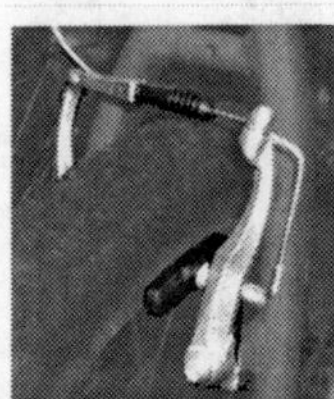

Linear-pull brake on rear wheel of a mountain bike

Linear-pull brakes or **direct-pull brakes**, commonly referred to by Shimano's trademark **V-brakes**, are a side-pull version of cantilever brakes and mount on the same frame bosses . However, the arms are longer, with the cable housing attached to one arm and the cable to the other. As the cable pulls against the housing the arms are drawn together. Because the housing enters from vertically above one arm yet force must be transmitted laterally between arms, the flexible housing is extended by a rigid tube with a 90° bend known as the "noodle". The noodle sits in a stirrup attached to the arm. A flexible bellows often covers the exposed cable.

Since there is no intervening mechanism between the cable and the arms, the design is called "direct-pull". And since the arms move the same distance that the cable moves with regard to its housing, the design is also called "linear-pull". The term "V-brake" is trademarked by Shimano and represents the most popular implementation of this design.

V-brakes usually lack an adjustment mechanism for cable length (other than the cable clamp and a barrel adjuster on the brake lever). They function well with the suspension systems found on many mountain bikes because they do not require a separate cable stop on the frame or fork. Because of the higher mechanical advantage of V-brakes, they require levers with longer cable travel than levers intended for calliper brakes or traditional cantilever brakes. This cable pull ratio was later adopted for disc brakes, making linear-pull levers standard for modern mountain bikes. Standard road levers normally do not pull enough cable to function with V-brakes.[4] To solve this problem, devices that use an eccentric pulley to increase the amount of cable pull of road levers, such as the "QBP Travel Agent", may be used.

Roller cam brakes

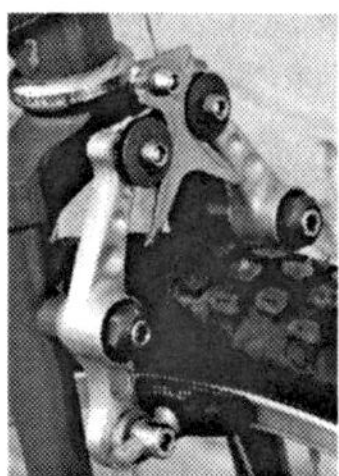

Roller cam front brake.

Roller cam brakes are centre-pull cantilever brakes actuated by the cable pulling a single two-sided sliding cam. (First and second-class lever designs exist; first-class is most common and is described here.) Each arm has a cam follower. As the cam presses against the follower it forces the arms apart. As the top of each arm moves outward, the brake shoe below the pivot is forced inward against the rim.[5] There is much in favor of the roller cam brake design. Since the cam controls the rate of closure, the clamping force can be made non-linear with the pull. And since the design can provide positive mechanical advantage, maximum clamping force can be higher than that of other types of brakes. They are known for being strong and controllable. On the downside, they require some skill to set up and can complicate wheel changes. And they require maintenance: like U-brakes, as the pad wears it strikes the rim higher; unless re-adjusted it can eventually contact the tyre's sidewall.

The roller cam design was first developed by Charlie Cunningham of WTB around 1982 and licensed to Suntour.[6] Roller cam brakes were used on early mountain bikes in the 1980s and into the 1990s, mounted to the head tube and seat stays in the standard locations, and below the chain stays for improved stiffness as they do not protrude to interfere with the cranks. It was not unusual for a bicycle to have a single roller cam brake combined with another type. They are still used on some BMX and recumbent bicycles.[7] Note that the common first-class lever roller cam brake is generally not convertible to second-class lever cantilever brake types as the pivots are in different locations.

There are two rare variants that use the roller cam principle. For locations where centre-pull is inappropriate, the side-pull **toggle cam brake** was developed.[8] Also a first-class cantilever, it uses a single-sided sliding cam (the toggle)[9] against one arm that is attached by a link to the other arm. As the cam presses against the (single) follower, the force is also transmitted to the other arm via the link. And specifically for suspension forks where the housing must terminate at the brake frame, the side-pull **sabre cam brake** was developed.[10] In the sabre cam design, the cable end is fixed and the housing moves the single-sided cam.

Delta brakes

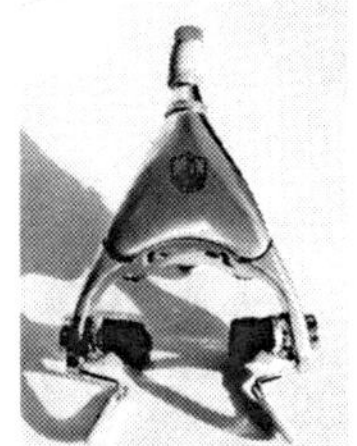

A Campagnolo delta brake.

The **delta brake** is a road bicycle brake named due to its triangular shape. The cable enters at the centre, pulls a corner of a parallelogram linkage housed inside the brake, pushing out the brake arms above the pivots, resulting in the arms below the pivots, with the pads, pushing in against the rim. While considered attractive, the delta brake has been criticized for being heavy, giving mediocre stopping power, and suffering disadvantageous variable mechanical advantage.[11] [12]

Made most prominently by Campagnolo in 1985, but also manufactured by Weinmann, and others,[13] they are no longer made and are now very uncommon.

Hydraulic rim brakes

Hydraulic rim brakes are one of the least common types. These brakes are generally able to be mounted on the same pivot points used for cantilever and linear-pull brakes. They were available on some high-end mountain bikes in the early 1990s, but declined in popularity with the rise of disc brakes. The moderate performance advantage (greater power and control) they offer over cable actuated rim brakes is offset by their greater weight and complexity. The only significant current use of these brakes is on bicycles used for trials riding.

Disc brakes

A hydraulic front disc brake, mounted to the fork and hub

Disc brakes consist of a metal disc attached to the wheel hub that rotates with the wheel. Callipers are attached to the frame or fork along with pads that squeeze together on the disc. Such brakes have been successfully used on motorcycles for decades, and are the principal choice there. They are finally becoming more popular on bicycles, after many (partly successful) attempts to introduce them over the last decades . Recent material advances in weight, costs and reliability have led several firms to develop and implement disc brake systems, and those are becoming a standard feature on many bicycles. They are used mainly on mountain bikes ridden off-road, but sometimes on hybrid bicycles and touring bicycles. Many tandem bicycles have a disc brake on the rear wheel in addition to rim brakes; the disc brake can be set to provide a constant drag, so that during long descents, the rim brakes are not overworked by the heavier machine.[14]

Advantages and disadvantages

Disc brakes perform equally well in all conditions including water, mud and snow. This is due to a number of factors:

- Their position closer to the hub and away from the ground and possible contaminants like water which can coat and freeze on the rim in colder temperatures keeps the disc rotor clean and working well.
- The disc brake calliper operates with a higher mechanical advantage than rim brakes, and as such squeeze the disc harder than rim brakes do the rim. For this reason, and due to the holes in the rotor, disc brakes maintain their stopping power in wet conditions, by more effectively clearing the disc of water.[15]
- Disc brakes are able to operate at a higher mechanical advantage than rim brakes because disc rotors in good condition are more true than rims in good condition, and thus do not need to retract as far from the rim when released.

They also avoid the problem that rim brakes have of wearing out the wheel rims, especially in muddy conditions, as well as the requirement that the rim be straight. The pads are usually made from metal sinters or an organic compound instead of rubber, and as such usually last longer than rim pads. Disc brakes offer better modulation of braking power and generally require less effort at the lever to achieve the same braking power. The use of tires as wide as 3.0 inches (76 mm) also makes disc brakes necessary, as rim brakes are not designed to straddle such a wide tire.

The benefits of disc brakes make them of great advantage in mountain bike riding, especially the more aggressive forms, such as freeride and downhill. Disc brakes are also becoming increasingly popular on hybrids, as they perform well in all weather conditions, and usually require less maintenance than rim brakes.

Disc brakes are sometimes heavier and more expensive than rim brakes, and require a hub built to accept the disc and a bicycle frame or fork built to accept the calliper. Older designs for front disc hubs often move the left hub's flange inward which causes the wheel to be dished, and therefore laterally weaker when forced to the non-disc side. Rigid forks on road bikes and tandems, made to handle the forces of a front disc brake, are heavier and may not have the ride quality of a regular fork.

A disc brake puts more stress on a wheel's spokes than a rim brake, since the torque of braking is between the hub and the rim. The spokes therefore must be stronger, this leads to slightly heavier and more expensive wheels.

The design and positioning of disc brakes can interfere with the pannier racks not designed for them. For this reason, many manufacturers produce "disk" and "non-disk" versions.

While a disc brake does not heat the rim, excessive heat build up can lead to disc failure. Bicycle discs are extremely lightweight, relative to the mass of bike and rider, compared to the large cast iron discs on an automobile, and therefore have little heat capacity. If brake friction exceeds convection and radiation losses, the temperature of the disc can very rapidly reach levels at which the metal begins to weaken and may warp or crack under braking stress. Therefore a disc is less suitable than a drum as a tandem drag brake, and heavy bikes (and riders) should choose the largest diameter discs available to increase heat capacity and cooling area.

Recently, a number of riders have experienced a dangerous problem with disc brakes. Under extreme braking conditions, the front wheel has come off the dropouts. Certain front forks using quick release skewers have been shown to have this problem. Riders should make sure the skewers are properly tightened before riding.[16]

Hydraulic *vs* mechanical

There are two main types of disc brake: mechanical (cable-actuated) and hydraulic. Mechanical disc brakes are almost always cheaper, but have less modulation, and may accumulate dirt in the cable lines since the cable is usually open to the outside.

Hydraulic disc brakes use fluid from a reservoir, pushed through a hose, to actuate the pistons in the disc calliper, that actuate the pads. They are better at excluding contaminants, but are difficult to repair on the trail, since they require fairly specialized tools. The brake lines occasionally require bleeding to remove air bubbles, whereas mechanical disc brakes rarely fail completely.

Also, the hydraulic fluid may boil on steep, continuous downhills. This is due to heat build up in the disc and pads and can cause the brake to lose its ability to transmit force ("brake fade") through incompressible fluids, since some of it has become a gas, which is compressible. To avoid this problem, 203 mm (8 inch) diameter disc rotors have become common on downhill bikes. Larger rotors require less calliper pressure for equal stopping power, dissipate heat more quickly, and have a larger amount of mass to absorb heat. Two types of brake fluid are used today: mineral oil and DOT fluid. Mineral oil is generally inert, while DOT is corrosive to frame paint but has a higher boiling point. Using the wrong fluid may cause the seals to swell or become corroded.

Single *vs* dual actuation

Many disc brakes have their pads actuated from both sides of the calliper, while some cheaper kinds have only one pad that moves. Many hydraulic disc brakes have a self-adjusting mechanism so as the brake pad wears, the pistons keep the distance from the pad to the disc consistent to maintain the same brake lever throw. Most mechanical discs have a manual control to adjust the pad-to-rotor gap. Callipers are now generally made in one piece to increase stiffness and reduce the threat of leaks, but the two-piece design still reduces heat build-up more effectively, and most top-end models still have a two-piece calliper. Also many top end callipers have four or more pistons as lower end models usually only have one or two.

Calliper mounting standards

There are many standards for mounting disc brake callipers. I.S. (International Standard) is different for 6-inch (150 mm) and 8-inch (200 mm) rotor and differs between forks with a QR and 20 mm through axle. The post-mount standard also differs by disc size and axle type. Many incompatible variants were produced over the years, mostly by fork manufactures. The mount used on the Rockshox Boxxer is the most typical of these specialty mounts, but most fork manufactures now use either the IS or post-mount standard for their current forks. As a point of reference, Hayes currently sells no less than 13 different adapters to fit their brakes to various mounting patterns.

Advantages and disadvantages of various types of mounts

A disadvantage of post mounts is that the bolt is threaded directly into the fork lowers. If the threading is stripped or if the bolt is stuck, then new fork lowers are required. Frame manufacturers have standardized the IS mount for the rear disc brake mount. In recent years post mount has gained ground and is becoming the most common. This is mostly due to decreased manufacturing and part cost for the brake callipers when using post mount . A limitation of the mount is that the location of the rotor disc is more tightly constrained: it is possible to encounter incompatible hub/fork combinations, where the rotor is out of range. With an IS mount, the calliper can be moved closer to or further from the mount point using spacers; this can permit a wider range.

Disc mounting standards

There are many options for disc rotor mounting - International Standard (IS), centerlock, Cannondale's 4-bolt pattern, Hope's 5-bolt pattern and Rohloff's 4-bolt pattern, to name a few. IS is a six-bolt mount and is the industry standard. Centerlock is patented by Shimano and uses a splined interface along with a lockring to secure the disc. The advantages of centerlock are that the splined interface is stiffer and removing the disc is quicker because it only requires one lockring to be removed. Some of the disadvantages are that the design is patented requiring a licensing fee from Shimano. A Shimano cassette lockring tool is needed to remove the rotor and is more expensive and less common than a Torx key. Advantages of IS six-bolt are that you have more choices when it comes to hubs and rotors. IS rotors use button head socket cap screws (typically M5x0.8x10mm with locking patch) with either a hex socket or Torx socket to secure them to the hub. This can make IS rotors more time consuming to remove. Torx screws are preferred for the superior torque: it is easy to strip the socket of a hex bolt by over tightening it, leaving a rotor that is hard to remove.

Standards

- International Standard (IS) (in widespread use)
- Centerlock (Shimano proprietary, in widespread use)
- Cannondale's 4-bolt pattern (obsolete)
- Rohloff's 4-bolt pattern (proprietary)
- Hope Technologies' 5-bolt pattern (Hope proprietary, obsolete)
- Hope Technologies' 3-bolt pattern (Hope proprietary)
- Rock Shox 3-bolt pattern (proprietary, obsolete)

Disc sizes

Disc brake rotors come in many different sizes, generally 160 millimeter, 185 mm, or 203 mm in diameter, however there are many different sizes available as all brake manufacturers make discs specific to their callipers and the dimensions often vary by a few millimeters. Larger rotors provide greater stopping power by virtue of a longer moment arm for the calliper to act on. Smaller rotors provide less stopping power but also less weight. Larger rotors will also dissipate heat more quickly preventing brake fade or failure. Typically downhill racers will run larger brakes to handle the greater braking loads and extended braking duration. Cross country racers will typically run smaller rotors which can easily handle the much smaller braking loads and offer a considerable weight savings of over 100g per rotor.[17] It is also common to use a larger diameter rotor on the front wheel and a smaller rotor on the rear wheel. This is due to the braking dynamics which shifts most of the rider weight to the front wheel during braking. This provides greater traction at the front wheel and allows for greater braking force. Conversely the weight shift off the rear wheel reduces its braking force. Using a smaller rear rotor will save weight and allow for better modulation of the rear brake while more efficiently using the wheel's braking capacity.

Drum brakes

Shimano Roller Brake unit on an internally geared hub.

Bicycle **drum brakes** operates like those of a car, although the bicycle variety use mechanical rather than hydraulic actuation. Two pads are pressed outward against the braking surface on the inside of the hub shell. Shell inside diameters on a bicycle drum brake are typically 70 – 120 mm. Drum brakes have been used on front hubs and hubs with both internal and external freewheels. Both cable- and rod-operated drum brake systems have been widely produced.

A **Roller brake** is a modular cable-operated drum brake for use on specially splined front and rear hubs. Unlike a traditional drum brake, the Roller Brake can be easily removed from the hub. It also contains a torque-limiting device called a power modulator designed to make it difficult to skid the wheel. In practice this can reduce its effectiveness on bicycles with adult-sized wheels.

Drag brakes are drum brakes intended to slow down a bicycle on long downhills rather than stop it. Drag brakes are employed on some tandem bicycles in mountainous areas where extended use of rim brakes can cause the rim to become hot enough to cause a blowout.[18] The largest manufacturer of this type of brake is Arai, whose brakes are screwed onto hubs with conventional freewheel threading on the left side of the rear hub and operated via Bowden cables.

Advantages and Disadvantages

Drum brakes are useful for wet or dirty conditions since the braking mechanism is fully enclosed. They are heavier, more complicated, and often weaker than rim brakes, but require less maintenance. They are most common on utility bicycles in some countries, especially the Netherlands, and are also often found on freight bicycles.

Coaster brakes

Cutaway view of a Husqvarna Novo coaster brake hub

The **coaster brake**, also known as a **back pedal brake** or **foot brake** (or **torpedo** in some countries), is a type of drum brake integrated into hubs with an internal freewheel. Freewheeling functions as with other systems, but when back pedaled, the brake engages after a fraction of a revolution. The coaster brake can be found in both single-speed and internally geared hubs.

When such a hub is pedaled forwards, the sprocket drives a screw which forces a clutch to move along the axle, driving the hub shell or gear assembly. When pedaling is reversed, the screw drives the clutch in the opposite direction, forcing it either between two brake pads and pressing them against the shell, or into a split collar and expanding it against the shell. The braking surface is often steel, and the braking element brass or phosphor-bronze, as in the Birmingham-made Perry Coaster Hub.

Coaster-brake bicycles are generally equipped with a single cog and chain wheel and often use 1/8" wide chain. However, there have been several models of coaster brake hubs with dérailleurs, most notably the Sachs 2x3. These use special extra-short dérailleurs which can stand up to the forces of being straightened out frequently and don't require an excessive amount of reverse pedal rotation before the brake engages. Coaster brakes have also been incorporated into hub gear designs - for example the AWC from Sturmey Archer, and the Shimano Nexus 3-speed.

Advantages and Disadvantages

Coaster brakes have the advantage of being protected from the elements and thus perform well in rain or snow. Though coaster brakes generally go years without needing maintenance, they are more complicated than rim brakes to repair if it becomes necessary. Coaster brakes also do not have sufficient heat dissipation for use on long descents. A coaster brake can only be applied when the cranks are reasonably level, limiting how quickly it can be applied. As coaster brakes are only made for rear wheels, they have the disadvantage common to all rear brakes of skidding the wheel easily. This disadvantage may, however, be alleviated if the bicycle also has a hand-lever-operated front brake and the cyclist uses it. As backpedaling is not possible with a coaster brake, it is necessary to place both feet on the ground, and then place the non-braking foot on the forward pedal to restart efficiently. A coaster brake is therefore not compatible with toe clips and straps, or with a clip-in shoe-pedal system.

Brake Levers

Brake levers are usually mounted on the handlebars within easy reach of the rider's hands. They may be distinct or integrated into the shifting mechanism. Road bicycles with drop handlebars may have more than one brake lever for each brake to facilitate braking from multiple hand positions.

Brake levers on the drop handlebars of a road bike with integrated shifters

Mechanical

Mechanical (cable) brake levers come in two varieties based on the amount of brake cable that they pull for a given amount of lever movement:

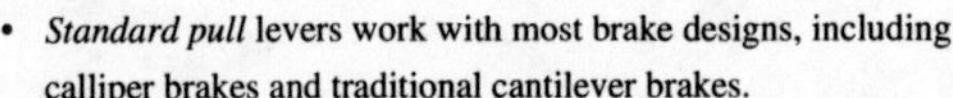

- *Standard pull* levers work with most brake designs, including calliper brakes and traditional cantilever brakes.
- *Long pull* levers work with "direct-pull" cantilever brakes, such as Shimano "V-Brakes"

The mechanical advantage of the brake lever must be matched to the brake it is connected to in order for the rider to have sufficient leverage to actuate the brake.

For example brake levers designed for calliper brakes may work with centre-pull cantilevers, but not with direct-pull, and linear-pull brakes. Direct pull cantilevers have twice as much mechanical advantage as traditional brakes, so they require a lever with half as much mechanical advantage. Long pull levers pull the cable twice as far, but only half as hard.[19]

Some bikes, especially road bikes with drop handlebars are set up with multiple levers to actuate a single brake. Levers that allow the rider to work the brakes from the tops of the bars, introduced in the 70s, are called *extension levers* or *safety levers*.[20] The modern equivalent are called *interrupt brake levers* and are considered superior.[21]

Hydraulic

Hydraulic brake levers push the hydraulic fluid down a hose towards the brakes.

Braking technique

There are several techniques for efficient braking on a standard, two-brake bicycle. The one most commonly taught is the 25-75 technique. This method entails supplying 75% of the stopping power to the front brake, and about 25% of the power to the rear. Since the bicycle's deceleration causes a transfer of weight to the front wheel, there is much more traction on the front wheel. Therefore, the rear brake can exert less braking force than on the front before the rear wheel starts skidding. For a more-detailed analysis, see Bicycle and motorcycle dynamics.

If too much power is applied to the front brake, then the momentum of the rider propels him/her over the handlebars, thereby flipping the bicycle. The skidding of the rear wheel can serve as a signal to reduce force on the front brake; a skillful cyclist in effect becomes a human anti-lock braking controller.

Some front brakes have a spring that limits the applied force; this is easier to use but limits the braking force and cannot compensate for changes in brake effectiveness due, for example, to a wet rim or overheated brake disc. On tandem bicycles and other long-wheel-base bicycles (including recumbents and other specialized bicycles), the lower relative centre of mass makes it virtually impossible for heavy front braking to flip the bicycle; the front wheel would skid first. On these bicycles, the safest quick stop is achieved with a somewhat higher proportion of force at the rear brake.

A skillful bicyclist often will use the front brake alone for moderate braking when riding on a good, paved surface. As the front wheel does not skid, the front brake poses less risk of loss of control, and does not cause rapid tire wear.

In some situations, it is advisable to slow down, and to use the rear brake more and the front brake less:

- When unfamiliar with the braking characteristics of a bicycle. It is important to test the brakes and learn how much hand force is needed when first riding a bicycle.
- When leaning in a turn (or preferably, brake before turning).
- Slippery surfaces, such as wet pavement, mud, snow, ice, or loose stones/gravel. It is difficult to recover from a front-wheel skid on a slippery surface, especially when leaned over.
- Bumpy surfaces: If the front wheel comes off the ground during braking, it will stop completely. Landing on a stopped front wheel with the brakes still applied is likely to cause the front wheel to skid and may flip the rider over the handlebar.
- Very loose surfaces (such as gravel and loose dirt): In some loose-surface situations, it may be beneficial to completely lock up the rear wheel in order to slow down or maintain control. On very steep slopes with loose surfaces where any braking will cause the wheel to skid, it can be better to maintain control of the bicycle by the rear-brake more than one would normally. However neither wheel should stop rotating completely, as this will result in very little control.
- Wet weather conditions, when the road surfaces are generally more slippery.
- Long descents: alternating the front and back brake can help prevent hand fatigue and overheating of the wheel rims which can cause a disastrous tire blow-out.
- Flat front tire: braking a tire that has little air can cause the tire to come off the rim, which is likely to cause a crash.[22]

It is customary to place the front brake lever on the left in right-side-driving countries, and vice versa,[23] because the hand on the side nearer the centre of the road is more commonly used for hand signals, and the rear brake can not pitch the bicyclist forward. However, a skilful bicyclist does better with the front brake on the side that is less often used for hand signals. In an emergency situation, operation of the brake has to be second nature; an unskilled bicyclist could find reversed brake levers confusing. Fortunately, it is usually easy to switch brake cables. The standard motorcycle configuration has the right-hand lever operating the front brake, and the left-hand lever operating the clutch. The levers are generally not easily reversed, leading to difficulties for a person who has learned on a bicycle with the front brake lever on the left.

Bicycles without brakes

→ Track bicycles are built without brakes so as to avoid sudden changes in speed when racing on a → velodrome. Since they have a fixed gear, braking can be accomplished by reversing the force on the pedals while the rider shifts their weight forward to accomplish a skid.

Some modern BMX bikes do not have brakes. The usual method of stopping is for the rider to put one or both feet on the ground, or to wedge a foot between the seat and the rear tire, but this can be very dangerous and is not recommended for riding on public highways or where a rider may collide with bystanders.

See also

- Bicycle and motorcycle dynamics
- Detangler
- Handle (grip)

References

[1] Hudson, William (2008). " Myths and Milestones in Bicycle Evolution (http://www.jimlangley.net/ride/bicyclehistorywh.html#tline)". Jim Langley. . Retrieved 2009-09-22.

[2] Rim brakes have been the subject of countless "engineering innovations". Some of the more unusual results can be seen here (http://www.blackbirdsf.org/brake_obscura/).

[3] Brown, Sheldon. " Adjusting Cantilever Brakes (http://sheldonbrown.com/canti-rollercam.html#u)". Sheldon Brown. . Retrieved 2009-07-31.

[4] http://www.sheldonbrown.com/canti-direct.html

[5] Pictures of a rear (http://www.flickr.com/photos/25373440@N00/280943857/sizes/o/in/set-72157602074286178/) and a front (http://www.flickr.com/photos/25373440@N00/280933087/sizes/o/in/set-72157602074286178/) roller cam brake.

[6] Private communication by Jeff Archer of the Museum of Mountain Bike Art & Technology (http://www.mombat.org)

[7] Brown, Sheldon. " Sheldon Brown's Bicycle Glossary (http://sheldonbrown.com/gloss_ri-z.html#rollercam)". Sheldon Brown. . Retrieved 2009-07-31.

[8] " Retrobike Gallery and Archive (http://www.retrobike.co.uk/forum/gallery2.php?g2_itemId=45595)". . Retrieved 2009-08-03. Drawings and technical description.

[9] An example of the toggle integral to the toggle cam brake appears here (http://flickr.com/photos/wombatbiker/2675719635/). The cam surface is the upper edge of the 'tail' on the large central piece. The cable attaches to one of the three holes. Note that the cam must flare at twice the rate of the two-sided roller cam design in order to move the arms the same amount.

[10] A picture of the rare sabre cam brake may be found at MOMBAT (http://www.mombat.org/1994_Phoenix.htm) - click on the 2nd thumb from the left.

[11] Heine, Jan (2008). " Slow Down, The Story of Bicycle Brakes (http://www.vintagebicyclepress.com/vbqindex.html)". *Bicycle Quarterly* (Vintage Bicycle Press LLC) (Winter 2008): 36. ISSN 1941-8809 (http://worldcat.org/issn/1941-8809). .

[12] Brandt, Jobst (October 2005). " Brakes from Skid Pads to V-brakes (http://www.sheldonbrown.com/brandt/brakes.html#delta)". sheldonbrown.com. . Retrieved 2009-01-22.

[13] Sutherland, Howard; et al. (1995). *Sutherlands Handbook for Bicycle Mechanics (6th Edition)*. Berkley, CA, USA. pp. 13.27 to 13.28. ISBN 0-914578-09-X.

[14] Brown, Sheldon. " Brakes for Tandem Bicycles (http://sheldonbrown.com/tandem-brakes.html)". Sheldon Brown. . Retrieved 2007-10-19.

[15] Heine, Jan (2008). "Slow Down, The Story of Bicycle Brakes". *Bicycle Quarterly* (Vintage Bicycle Press LLC) (Winter 2008): 39. ISSN 1941-8809 (http://worldcat.org/issn/1941-8809).

[16] Annan, James. " Disk brakes and quick releases - what you need to know (http://www.ne.jp/asahi/julesandjames/home/disk_and_quick_release/)". . Retrieved 2007-10-19.

[17] " Disc Brake weight listing (http://weightweenies.starbike.com/listings/components.php?type=discbrakes)". . Retrieved 2006-11-07.

[18] " Sheldon Brown Glossary: Drag Brake (http://www.sheldonbrown.com/gloss_dr-z.html)". . Retrieved 2008-05-20.

[19] Brown, Sheldon. " Sheldon Brown Glossary: Brake (http://sheldonbrown.com/gloss_bo-z.html#brake)". Sheldon Brown. . Retrieved 2008-02-01.

[20] " Sheldon Brown's Glossary: Extension levers (http://www.sheldonbrown.com/gloss_e-f.html#extensionlevers)". . Retrieved 2009-09-23.

[21] " Sheldon Brown's Glossary: Interrupter Brake Levers (http://www.sheldonbrown.com/gloss_i-k.html#interrupter)". . Retrieved 2009-09-23.

[22] Brown, Sheldon. " Braking and Turning your Bicycle (http://www.sheldonbrown.com/brakturn.html)". Sheldon Brown. . Retrieved 2007-10-19.
[23] Brown, Sheldon. " Cables: Right front or left front? (http://www.sheldonbrown.com/cables.html)". Sheldon Brown. . Retrieved 2009-08-06.

- Bicycle Glossary (http://sheldonbrown.com/glossary.html) from Sheldon Brown's website
- **(Swedish)** Ekström, Gert; Husberg, Ola (2001) (in Swedish). *Älskade cykel* (1st ed.). Bokförlaget Prisma. ISBN 91-518-3906-7.

List of bicycle parts

List of bicycle parts

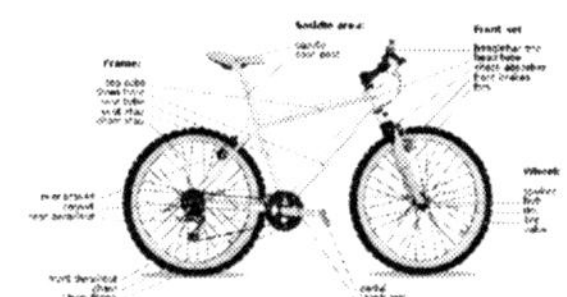

Bicycle parts

- **Axle** - as in the generic definition, a rod that serves to attach a wheel to a bicycle and provides support for bearings on which the wheel rotates. Also sometimes used to describe suspension components, for example a swing arm pivot axle.
- **Bar ends** - extensions at the end of straight handlebars to allow for multiple hand positions.
- **Bar plugs** aka end caps - Handlebar plug - plugs for the ends of handlebars.
- **Bearing** - a device that facilitates rotation by reducing friction. The most common types are ball, roller and sleeve.
- **Belt-drive** - alternate to chain-drive
- **Basket** - cargo carrier
- **Bottle cage** - a holder for a water bottle
- **Bottom bracket** - The bearing system that the pedals (and cranks) rotate around. Contains a spindle to which the crankset is attached and the bearings themselves. There is a bearing surface on the spindle, and ones on the cups that thread into the frame. The bottom bracket may come be overhaulable (an adjustable bottom bracket) or not (a cartridge bottom bracket). The bottom bracket fits inside the bottom bracket shell, which is part of the bicycle frame.
- → **Brake** - Brakes are used to stop the bicycle. *Rim brakes* and *disc brakes* are operated by brake levers, which are mounted on the handlebars. *Coaster brakes* are operated by pedaling backward.
- **Braze-on** - a fitting protruding from a frame to provide attachment, typically for cable housings or tire pumps and similar accessories.
- **Cable guide** - A fitting below the bottom bracket which guides a piece of bare inner bowden cable around a corner.
- **Cable** - a metal cable enclosed in part by a metal and plastic housing that is used to connect a control, such as a brake or shifting lever, to the device it activates.
- **Cartridge bearing** - A type of bearing that is not user-serviceable, but must be replaced as a unit.
- **Cassette** - a group of stacked sprockets on the rear wheel of a bicycle with a rear derailleur.
- **Chain** - A system of interlinking pins, plates and rollers that transmits power from the front cranks to the rear wheel.
- **Chainring** - (one of the) front gear(s), attached to a crank arm.
- **Chainstay** - pair of tubes on a bicycle frame that runs from the bottom bracket to the rear dropouts
- **Cogset** - the set of rear sprockets that attaches to the hub on the rear wheel.
- **Cone** - holds bearings in place, pressed against the cup
- **Crankset** - composed of crankarms and chainrings

- **Cotter** - pin for attaching cottered crank arms
- **Coupler** - to connect tubing together
- **Cup** - receives ball bearings which roll along its inner surface; integrated on most conventional hubs or can be pressed into older bottom bracket shells. *See also: Cone*
- **Cyclocomputer** - an electronic accessory that measures and displays instantaneous and cumulative speed and distance. Often provides other measurements such as heart rate.
- **Derailleur hanger** - a piece on the rear dropout that the derailleur attaches to.
- **Derailleur** - an assembly of levers, usually cable actuated, that moves the chain between sprockets on a cassette or chainring assembly.
- **Down tube** - tube on a bicycle frame that runs from the head tube to the bottom bracket.
- **Dropout (bicycle part)** - paired slots on a fork or frame at which the axle of the wheel is attached
- **Dustcap** - any cap serving to keep dirt and contamination out of an assembly. Common over crank bolts, often plastic.
- **Dynamo** - bicycle lighting component, aka generator.
- **Eyelet** 1. attachment point on frame, fork, or dropout for fenders, racks, etc.

  2. a hole through which a spoke nipple passes through the rim so it may attach to a spoke
- **Fairing**
- **Fender**
- **Ferrule** a metal or plastic sleeve used to terminate the end of a cable housing
- **Fork** - a mechanical assembly that integrates a bicycle's frame to its front wheel and handlebars, allowing steering by virtue of its steerer tube.
- **Frame** - the mechanical core of a bicycle, the frame provides points of attachment for the various components that make up the machine. The term is variously construed, and can refer to the base section, always including the bottom bracket, or to base frame, fork, and suspension components such as a shock absorber.
- **Freehub** - a ratcheting assembly onto which a cog or cassette is mounted that allows the bicycle to coast without the pedals turning.
- **Freewheel** - a ratcheting assembly that that incorporates cogs and that allows the bicycle to coast without the pedals turning.
- **Hanger** - part of frame or an attachment to the frame to which the derailleur is attached (see Derailleur hanger)
- **Handlebar tape** - a tape wound around dropped handlebars so as to provide padding and grip, usually cork or cloth, sometimes foam rubber.
- **Handlebar** - a lever attached, usually using an intermediary stem, to the steerer tube of the fork. Allows steering and provides a point of attachment for controls and accessories.
- **Head badge** - manufacturer's or brand logo affixed to the head tube
- **Head tube** - the tube of a bicycle frame that contains the headset
- **Headset or head set** - the bearings that form the interface between the frame and fork steerer tube
- **Hood** - The rubber brake lever covering on bikes with drop style handle bars
- **Hub** - the core of a wheel - contains bearings and, in a traditional wheel, has drilled flanges for attachment of spokes.
- **Hub dynamo** - a generator inside one of the hubs for powering lights or other accessories
- **Indicator** - a turn signal - see Bicycle lighting#Turn signals (indicators)
- **Inner tube** - a bladder that contains air to inflate a tire. Has a Schrader or Presta valve for inflation and deflation.
- **Jockey wheel** - one of two small sprockets of the rear derailleur that guide the chain
- **Kickstand** - a folding attachment used to park a bicycle upright. Usually mounts to frame near bottom bracket, sometimes near rear dropouts.
- **"Lawyer tab"** - also called "lawyer lips", a retention device on the dropouts of the front fork to prevent inadvertent loss of the front wheel in the case it is not properly secured.

- **Locknut** - a nut designed not to loosen due to vibration.
- **Lockring** - a ring, usually metal, of varying design, that serves to retain a component in place.
- **Lug** - a metal connector used to align frame components where they join each other.
- **Luggage carrier** - any accessory equipment designed to carry tools, gear or cargo.
- **Nipple** - a specialized nut that most commonly attaches a spoke to a wheel rim. In some systems, it provides attachment to the hub.
- **Pannier** - cloth zippered storage bags that mount to sides of luggage racks.
- **Pedal** - mechanical interface between foot and crank arm. There are two general types - one secures the foot with a mechanical clamp or cage and the other has no connection to lock the foot to the pedal.
- **Quick release** - a skewer with a lever on one end that loosens when the lever is flipped. Used for releasing wheels and seat posts.
- **Rack** - a rack that attaches behind the seat, usually with stays to the rear dropouts, that serves as a general carrier.
- **Reflector** - reflects light to make bicycle evident when illuminated by headlights of other vehicles. Usually required by law but held in disdain by many cyclists.
- **Rim**
- **Rotor** - a device that allows the handlebars and fork to revolve indefinitely without tangling the rear brake cable
- **Safety levers**, extension levers, and interrupt brake levers
- **Saddle** - also seat. What you sit on.
- **Seat** - also saddle. What you sit on.
- **Seat Rails** - a metal framework over which saddle covering is stretched. The seat post attaches to the seat rails by means of a clamp.
- **Seat lug** - a frame lug on the top of the seat tube serving as a point of attachment for a clamp to secure the seat post.
- **Seat tube** - the roughly vertical tube in a bicycle frame running from the seat to the bottom bracket.
- **Seat bag** - a small storage accessory hung from the back of a seat.
- **Seatpost** - a post that the seat is mounted to. It slides into the frame's seat tube and is used to adjust ride height depending how far into the seat tube it is inserted.
- **Seatstay** - frame components, small diameter tubes running from top of seat tube to rear dropouts.
- **Shaft-drive** - alternate to chain-drive
- **Shifter** - see also Ergo Shifting and Shimano Total Integration, two competing methods of combined shifter and brake lever controls
- **Shock absorber** - for bicycles with suspensions, a device that limits the rate at which suspension rebounds after absorbing an impact.
- **Skirt guard** or coatguard - a device fitted over the rear wheel of a bicycle to prevent a long skirt, coat or other trailing clothes or luggage from catching in the wheel, or in the gap between the rim and the brakes.
- **Spindle** - an axle around which a pedal rotates - threaded at one end to screw into crank arms.
- **Spoke** - connects wheel rim to hub. Usually wire with one end swaged to form a head and one threaded end. A typical wheel has 36 spokes.
- **Steering tube** - a tube on top of a fork that is inserted through frame and serves as an axle by means of which bicycle can be steered.
- **Stem** - a bracket used to attach handlebars to steerer tube of fork. Usually secured by pinch bolts.
- **Tire** - as in common usage. Usually pneumatic. A tubular tire is glued to the wheel rim; most tires use tubes, but tubeless tires and rims are increasingly common.
- **Toe clips** - a metal or plastic cage attached to a pedal. Usually has an adjustment strap. Secures foot to pedal for increased control and more effective transfer of power from foot to drive chain.
- **Top tube** - frame member leading from steerer tube to seat tube.

- **Valve stem** or simply **valve** - port for adding or releasing air from the inner tube. Two types are commonly used: Presta and Schrader. A third type, the Woods or Dunlop valve, can still be found in Europe and Asia.
- **Wheel** - as in common usage. Traditionally and most commonly spoked.

References

- International Standard ISO 8090: Cycles — Terminology(same as: British Standard BS 6102-4).

See also

- Groupset
- Bicycle tools

External links

- Multilingual Bicycle's Vocabulary [1]

References

[1] http://bemi.free.fr/biciklo/

Track cycling

Track cycling is a bicycle racing sport usually held on specially-built banked tracks or → velodromes (but many events are held at older velodromes where the track banking is relatively shallow) using → track bicycles.

Track racing is also done on grass tracks marked out on flat sportsfields. Such events are particularly common during the summer in Scotland at Highland Games gatherings, but there are also regular summer events in England.

Riding position

A Track Cycling Race

Aerodynamic drag is a significant factor in both road and track racing. Frames are often one-piece molded carbon fiber affairs which allows a lightweight and aerodynamically "slippery" design. More traditional bikes might employ airfoil cross sectional shapes in the frame tubes and ever greater attention is being paid to aerodynamics in component group design.

Given the importance of aerodynamics, the riders' sitting position becomes extremely important. Handlebars on track bikes used for longer events such as the points race are similar to the drop bars found on road bicycles. The riding position is also similar to the road racing position.

In the sprint event the rider's position is more extreme compared with a road rider. The bars are lower and the saddle is higher and more forward. Bars are often narrower with a deeper drop. Steel bars, as opposed to lighter alloys or carbon fiber, are still used by many sprinters for their higher rigidity and durability.

In timed events such as the pursuit and the kilo, riders often use aerobars or 'triathlon bars' similar to those found on road time trial bicycles, allowing the rider to position the arms closer together in front of the body. This results in a more horizontal back and presents the minimum frontal area to reduce drag. Aerobars can be separate bars that are attached to time trial or bull horn bars, or they can be part of a one-piece monocoque design. Use of aerobars is permitted only in pursuit and time trial events.

Formats of track cycle races are also heavily influenced by aerodynamics. If one rider closely follows, they draft or slipstream another, because the leading rider pushes air around themselves; any rider closely following has to push out less air than the lead rider and thus can travel at the same speed while expending less effort. This fact has led to a variety of racing styles that allow clever riders or teams to exploit this tactical advantage, as well as formats that simply test strength, speed and endurance.

During the early 1990s in individual pursuit events, some riders, most notably Graeme Obree, adopted a straight-armed *Superman*-like position with their arms fully extended horizontally, but this position was subsequently outlawed by the Union Cycliste Internationale, the sport's ruling body. Recumbent bicycles can actually be ridden faster, but are banned from UCI competition. The International Human Powered Vehicle Association is a separate organisation that runs recumbent races, including the human-powered speed record.

Main centres

Track cycling is particularly popular in Europe, notably Belgium, France, Germany and the United Kingdom where it is often used as off-season training by road racers (professional six-day 'Madison' events were often entered by two-man teams comprising a leading road racer and a track specialist).

The sport also has significant followings in Japan and Australia. It is part of the Summer Olympic Games, and there are UCI Track World Championships as well as circuits of professional events in many areas.

In the United States, track racing reached a peak of popularity in the 1930s when six-day races were held in Madison Square Garden in New York. The word "Madison" is still used as the name for this type of race in six-day racing.

Race Formats

Track cycling events fit into two broad categories, Sprint races and Endurance races. Riders will fall into one category and not compete in the other. Riders with good all round ability in the junior ranks will decide to focus on one area or another before moving up to the senior ranks.

Sprint races are generally between 3 and 8 laps in length and focus on raw sprinting power over a small number of laps and race tactics to defeat opponents. Sprint riders will train specifically to compete in races of this length and will not compete in longer endurance races. Famous Sprint track riders at present include Chris Hoy and Jason Kenny of Great Britain and Theo Bos of The Netherlands.

Main Sprint Events

- Sprint
- Team sprint
- Keirin
- Track time trial

Endurance races are held over much longer distances. While these primarily test the riders endurance abilities, the ability to sprint effectively is also required in the Madison, Points Race and Strach Race. The length of these races varies from 12 - 16 laps for the Individual and Team Pursuit races, right up to 200 laps for a full length Madison race in World Championships or Olympic Games.

Main Endurance Events

- Individual pursuit
- Team pursuit
- Scratch race
- Points race
- Madison

- Omnium
- Handicap
- Miss and Out, elimination or 'Devil Take the Hindmost'

World Track Cycling - Major Events

Olympic Games

Held every four years as part of the summer Olympics. There are currently 10 events in the Olympics, less than appear in the World Championship. 7 of these events are for men while only 3 are for women. Competition is held over five days.

As with other Olympic events the winner of each event is Olympic Champion and gold medal winner, while second and third places receive silver and bronze. At the most recent Olympics in China in 2008, Great Britain were the most successful nation in track cycling. They won 7 out of the ten events and also won several silver and bronze medals. Chris Hoy won three gold medals while Bradley Wiggins won two. The next summer Olympics are in London in 2012.

UCI Track World Championships

Held every year, usually in March or April at the end of the winter track season. There are currently 17 events in the World Championships, 9 for men and 8 for women. Qualification places are determined by different countries performance during the World Cup Classic series held through the season (see below).

The winner of each event wins the title of world champion and the gold medal. They are presented with the world champions rainbow jersey. This is a white jersey with rainbow stripes across the centre and can be won with pride by the winner whenever they compete in that event over the coming year. Second and third placed riders win the silver and bronze medals. The most recent World Championship were held in Pruszków in March 2009, the 2009 UCI Track Cycling World Championships. Australia finished as the most successful nation in these championships. They won 4 of the 19 events and also several silver and bronze medals. Grégory Baugé and Alex Rasmussen both took two world titles while Elizabeth Armitstead, Simona Krupeckaitė and Victoria Pendleton won a total of three medals each. The next championships will be held in Ballerup, Denmark in March 2010.

UCI Track Cycling World Cup Classics

The World Cup Classics series consists of 5 events for 2008, previously 4, held in different countries throughout the world during the winter track cycling season. These meeting include all of the 17 events that take place in a World Championship and are usually spread over three days.

Events won and points scored by the riders throughout this series count towards qualification places for their nation in the World Championships at the end of the season. The overall leader in each event wears the points leaders jersey at each race, with the overall winner at the end of the season keeping the jersey and wearing it at the World Championships. Riders compete for their own country or as part of a sponsored trade team at these events. Therefore it is possible for a number of teams from one country to compete at each event.

As World Championship qualification is at stake, the events do attract a top field of riders. However it is common for top riders not to compete at all the events of the series, with teams/countries often using the events to field younger riders or attempt different line ups at some events. Top riders can still win the series, or obtain good a placing for qualification points for their country, without competing at every event.

The first event of the 2008/2009 season took place in Manchester in October/November 2008. The event, on the back of Britain's recent World Championship and Olympic success, was a complete sell out. Great Britain dominated this meeting, winning 14 of the 17 events.

The first event of the 2009/2010 season will again be in Manchester, from 30 October to 1 November. The event's website is www.trackworldcup.co.uk

Track records

In addition to regular track racing, tracks are also the venue for many cycling records. These are over either a fixed distance or for a fixed period of time. The most famous of these is the hour record, which involves simply riding as far as possible in one hour. The history of the hour record is replete with exploits by some of the greatest names in cycling from both road and track racing (including, among others, Major Taylor, Henri Desgrange, Fausto Coppi, Jacques Anquetil, Eddy Merckx, Francesco Moser, Miguel Indurain and Tony Rominger). Originally, attempts were made at velodromes with reputations for being fast (such as the Velodromo Vigorelli in Milan). More recently, attempts have moved to high-altitude locations, such as Mexico City, where the thinner air results in lower aerodynamic drag, which more than offsets the added difficulty of breathing. Innovations in equipment and the rider's position on the bike have also led to dramatic improvements in the hour record, but have also been a source of controversy (see Graeme Obree).

See also

- Union Cycliste Internationale
- Schuermann
- → Velodrome
- List of cycling tracks and velodromes
- Six-day racing

External links

- The Velodrome (www.velodrome.org.uk), A home for Track Cycling on the web [1]
- FixedGearFever (www.FixedGearFever.com), The Virtual Velodrome - the largest track cycling community on the net! [2]
- TrackCyclingSA.co.za - South Africa's first dedicated track cycling site [3]
- USA Cycling - Track [4]
- British Cycling track news and information site [5]
- Number one source for Track Equipment, Keirin Cycle Culture Cafe, Berlin [6]
- American Track Racing Association (ATRA) - Formed by United State and Canadian velodromes to advance the sport of track cycling, the capital improvements of their facilities and for the promotion of track riding and racing. [7]

References

[1] http://www.velodrome.org.uk
[2] http://www.fixedgearfever.com
[3] http://www.TrackCyclingSA.co.za
[4] http://www.usacycling.org/track/
[5] http://www.britishcycling.org.uk/web/site/BC/tra/track_latest_news.asp
[6] http://www.keirinberlin.de
[7] http://raceatra.com/

Article Sources and Contributors

Fixed-gear bicycle *Source*: http://en.wikipedia.org/w/index.php?title=Fixed-gear_bicycle *Contributors*: 842U, Ajdelore, Ancos, AndrewDressel, AndyArmstrong, AndyMorris, Animal Lairs, Apefish, Arundelo, Avijja, Banaticus, Barticus88, Beach drifter, Beastah89, Bguenther, BitQuirky, Bklingenberg, BruceMcAdam, Bryan Derksen, Buffalo Bill, Cacophony, Camdiamond, Camerhogneman, Cbass94, Ccastill, Cdc, Ceehood1, Charliethebikemonger, Chinasaur, Christopherlin, Cosmo the third, DTM, Danielfolsom, Danielhorner@mail.com, Dhodges, Djharrity, Doczilla, Drewish, Ds13, Elipongo, Emanuelbri, EoGuy, Exidor, Existentialmedia, Falc0n2600, Fenwayguy, Fixedgearfever, Fletcher, Fltchr, Frymaster, Gary Cziko, George The Dragon, Gogo Dodo, GoneAwayNowAndRetired, Goworldgo, Grenavitar, Hartleymartin, Interiot, Ironman1104, Itsi7alian, JaGa, Jason McConnell-Leech, Javert, JeffreyHolliday, John Nevard, John Reaves, Joi, Julius.kusuma, Jumble, Kalliona, Keithonearth, Kouban, Kozuch, Kurtdriver, LDHan, Lesnail, Lheydon, Luna Santin, MLRoach, Manderr, Mani1, Mannybalalio, Marek69, Mattoondah, Mboverload, Mipadi, Moriquendi, Mushroom, Musicbulldog, NickBush24, Nixeagle, Nmwarren1, One Night In Hackney, Piano non troppo, Plasticup, Plebeian, ProfDEH, Proofreader77, Qartis, Radiokillplay, Reisio, RepublicanJacobite, ReyBrujo, Rjwilmsi, Rogerzilla, Rwoodsmall, S2s2steven, Sdrtirs, Severo, SiobhanHansa, SkonesMickLoud, Snori, Stephan Leeds, Stephen Riley, Sudbrancxeto, Suleyman Habeeb, Swassafrass, Tabledhote, Taisau, The Anome, The Thing That Should Not Be, The Ungovernable Force, Theendlessdream, Thewalrus, Tide rolls, Tiggerjay, Timwi, Tyler69, Unregistered.coward, Useight, Viasolus, Vicioustwist, WAvegetarian, Will Beback, Will.law, WorldDownInFire, Zer0Nin3r, 340 anonymous edits

Bicycle *Source*: http://en.wikipedia.org/w/index.php?title=Bicycle *Contributors*: (aeropagitica), 050101 manly24, 1-is-blue, 119, 16@r, 21655, 270408j, 2D, 3dnatureguy, 3rdsanekid, 3xcrazy, 4I7.4I7, 69mehard, A3RO, ABF, AHands, Aaa112, Aarond144, Abce2, Abductive, Abdullais4u, Academic Challenger, Acroterion, Adam78, AdamRetchless, Adamrice, Adashiel, Addshore, AdjustShift, Adolphus79, Ae-a, Afri, Agathman, Ahoerstemeier, Ahunt, Airbete, Aitias, Ajentp, Akamad, Aksi great, AlainV, Alan Canon, Alan Liefting, Alansohn, Alatius, Alex LaPointe, Alex Ramon, Alistairseed, Allen3, Allstarecho, Alphachimp, Alterego, Alureiter, American McBastard, Aminalshmu, Amore Mio, Anaxial, Anclation, AndonicO, Andre Engels, Andres, Andrew Nutter, AndrewDressel, Andrewpmk, Andries, Andromeda321, Andy Richter, Anetode, Angela, Angr, Animum, Anna Lincoln, AnnaFrance, Annoyingdevilofbeep, Antandrus, Anthony, Antitext, Antonio Lopez, Aoljimmy, Ap, Apbiologyrocks, Apparition11, Applepies66, ApplesauceABC123, Aratuk, Arienh4, Arjun01, Arnon Chaffin, Art LaPella, Arteitle, Arthal, Arthena, Ascidian, Ashley Pomeroy, Asirota, AstroHurricane001, Atlant, Atomictomato, Atxo, Avono, Awesomex2, Awien, Azxten, B, BRG, Badgernet, Badmonkey, Bakabaka, Balcer, Bardok55, Barrylb, Bateman1234, Bcorr, Bdavisongsu, Bdwolverine87, BearPer, Bearcat, Behun, Bejinhan, Beland, Ben48, Bensifey, Beringar, Berria, Besselfunctions, Bestbrandworld, Bicycle repairman, Big Bird, BigFatBuddha, Bigbrainsam, Bige1977, Bihco, Bikeduckquack, Bikepower, BillC, Bitbucket, BjKa, Bjornc, BlaiseFEgan, Blake-, Blankinfinity, BlastOButter42, Bletch, Blimpguy, Blondeonblonde, Blue Order, Bluelion, Bmxtl32, Bncbrowne, Bo0608, Bob Burkhardt, Bob91919, Bobblewik, Bobjojim, Bobo192, Bobvillanova, Bogey97, Bonadea, Bongomatic, Bongwarrior, Boobs84, Born2cycle, BradBeattie, Bradley2946, BrainyBabe, Bravixaton, BrianKnez, Brianary, Brianga, Brilliant trees, BrokenSegue, Brycen, Bsadowski1, Bumhole123, Bwileyr, C.thompson, CALR, CO, CWY2190, CWii, CZmarlin, Calamarain, Caltas, Camaron, Cambrasa, Camphost, Can't sleep, clown will eat me, Canaima, Canderson7, Canislupusarctos, Capricorn42, Captain panda, Captainpeejay, CardinalDan, CarolGray, Cassandra 73, Cataclysm, Catbar, Catherineyronwode, Ccaravello, Cdamian, Cdc, Cdills, Celithemis, Cflm001, CharlesC, Chasingsol, Chicbicyclist, ChiragPatnaik, Choda, Choda 2, Chris 73, Chris.skelding92, ChrisHamburg, Chrisch, Chrishans, Christopherlin, Chuanha1, Churchh, Ckatz, Clarkquest, Cloktox, ClosedEyesSeeing, Closedmouth, Cmichael, Cnsupplier, Cobaltbluetony, Codetiger, Coemgenus, Coffee, Colipon, Comgoog, Commander, CommonsDelinker, Conversion script, Cookie90, Cool3, CoolGuy, Coralmizu, Cosmicosmo, Countspiw3, Cpcheung, Crashman alpha, Crazysuit, Crissov, Crowstar, Crumbsucker, Cubs Fan, Cucuheader, Cutler, Cyanidethistles, CyclePat, D3v4st4t0r, DGJM, DHN, DJ Clayworth, Damian Yerrick, Dan D. Ric, Dancarney, Dancter, Dandelion1, DanielLeeRoss, DannyWilde, Darkxavier, Darwinek, David Gale, David.Mestel, DavidJavidzad, Davids5566, Dbfirs, Dcoetzee, DeFacto, DeadEyeArrow, Declan Clam, Decltype, Deconstructhis, Delldot, DerHexer, Dfrg.msc, Dhodges, Dieter Weber, Diomidis Spinellis, Dirkbb, DirkvdM, Dischdeldritsch, Discospinster, Dismas, Dizzydog11235, Djetelina, Dmsar, Doc Tropics, Donarreiskoffer, Dostal, Douglaz150, Doulos Christos, Downwards, Dr. Marcus Gossler, Dr.Malooden, Drafzilla, Drake Wilson, Drbug, Dreadstar, Drini, Drj, Drmaik, Dromioofephesus, Drunkenmonkey, Dspradau, Dudley360, Dutchbikes, Dyamantese, Dyl, Dynaflow, Dysepsion, E. Ripley, ERcheck, ESkog, EWS23, Eamonnca1, Ed Fitzgerald, EdX20, Ee43, Eep², Eggoeater, Egil, Ehteshami, Elagatis, Elassint, Elephant53, Eliashc, Eliharold, Elipongo, Elkman, Elm-39, Emijrp, Emokid4000, Empirecontact, Enchanteddrmzceo, Ender qa, Ensiform, Enviroboy, Epbr123, Eranb, ErgoSum88, Eric-Wester, Erik Kok, Escape Orbit, Euchiasmus, Evaluist, Everyking, Excirial, Farmer21, FastLizard4, Fastfission, Fat and Bald, FayssalF, Fayylard, Feezo, FelisLeo, Fifelfoo, Figma, Filelakeshoe, Finchsnows, Fingal, FisherQueen, Flowerparty, Flyguy649, Footballfan190, Forbsey, Forwardxo, Fracaxxo, Frankie0607, Freakofnurture, Frederock, Fredfred2, Fredrik, Freodeo, FreplySpang, Fruitrocks1000x, Fulmar2, Funnybunny, G, G-Man, GB fan, GDonato, GRAHAMUK, Gaaras girl2, Gadfium, Gail, Gaius Cornelius, Galoubet, Gamaliel, Gary Cziko, Gawaxay, Gdr, Geg, Gengiskanhg, Geoave, Geobio, Georgette2, GeraldTehLegend42, Gesslein, Gfox88, Ggisalegend, Ghiraddje, Giantsfan1, GilbertoSilvaFan, Gilliam, Gimboid13, Gmaxwell, Gnewf, Gobeirne, Godlord2, Gogo Dodo, Goodnightmush, Gopats1221, Gordonrox24, Gothicemo45, GraemeL, Graffitiisart, Graham Bogdanov, Graham87, Greatwalk, Green commuter, Gregfitzy, Gregorydavid, Grenavitar, Gromlakh, Grove-m8, Grstain, Gruk, Grunt, Gtstricky, Gun Powder Ma, Gurch, Gwernol, Gzkn, Gzornenplatz, Gzuckier, H3mcostlymilk, HEL, Haabet, Hadal, Hairy Dude, Hank chapot, Hard12beat, Hari, Harrigan, Harry491, Hasek is the best, Hashlete, Hbent, Hdpinn, HeikoEvermann, Hektor, Hellomensd, Hemanshu, Henningklevjer, Heron, HexaChord, Hiteshnairhitesh, Hitsnthx, Hobartimus, Hobofett12, Holeymoleytoad102, Hopefullynot, Horsten, Hotlorp, Howcheng, Hu12, Hughhunt, HumphUK, Husond, Huy96, II MusLiM HyBRiD II, IRP, Ian Moyes, IanHarvey, IanOsgood, Ibasket, Iijjccoo, Ilya, IndianCow, Ingolfson, Inkington, InvisibleK, Ipatrol, Iridescent, Issam1, Ivysaur, Ixfd64, J-rod, J.delanoy, JBellis, JCblows, JFreeman, JNW, JSpung, JaGa, Jacbat, Jackol, Jacobolus, Jaja123456, Jaknouse, Jan eissfeldt, Janke, Jauerback, JavierMC, Jayp1981, Jbolden1517, Jcvamp, Jdorwin, Jeff G., Jennavecia, Jens Schriver, JethroElfman, JimRaynor, Jin Man Choi, Jj137, Jkrafcik, Jmundo, JoanneB, Joelr31, Johann Wolfgang, John, John Reaves, JohnMcDonnell, Jojhutton, Jon Harald Søby, Jonemerson, Jons63, Jorfer, Jose77, Josephataya, Joshgraham, Joy, Joyous!, JudahH, Julius.kusuma, Jusjih, Just zis Guy, you know?, Jwanders, Jwkpiano1, JzG, KJS77, KSava, Kafziel, Kaobear, Katalaveno, Katefan0, Katieh5584, Kbutler80, Kchishol1970, Kcp0768, Keilana, Keinstein, Keithonearth, Kelvinc, Kevin Saff, Kevinti, Kgasso, Khalidkhoso, Khashmi316, Khukri, Kianabc, Kickstart70, Kidblast, Killiondude, Kilonum, KindGoat, Kingandrew, Kingkebab, Kingpin13, Kirchsw, Kirtai, Kizor, KnowledgeOfSelf, Koavf, Kotra, Kotts, Kozuch, Krich, Kristen Eriksen, Kukini, Kuru, KypDurron1, LDHan, La Pianista, Lankiveil, Larsobrien, Law, LeadSongDog, LeaveSleaves, Legend, Lenticel, Leo44, Leonard G., Les woodland, Lesseps, Leujohn, Leuqarte, LibLord, Liface, Liftarn, Lightmouse, Lights, Lilx, LittleDan, LittleOldMe, Livajo, LizardJr8, Londo06, LondonStatto, Lord Emsworth, Loren.wilton, Lorinda, LovesMacs, Lucasbfr, Luna Santin, Luxoran, M1ss1ontomars2k4, MBisanz, MER-C, MPF, Mabisa, Mac, Macaddct1984, MadCowDisease675, MadFunk, Magnus, Magnuschubb, Mahjongg, Mailer diablo, Mais oui!, Makemi, Man vyi, Mandarax, Manderr, Mani1, Marek69, Markus Kuhn, Maros, Martijn faassen, Martin451, MarylandArtLover, Maschwab, Masteralex007, Mastrchf91, Matt Deres, Matt me, MattIndustries, Mattaber, Mattyew, Maury Markowitz, Maximus Rex, Mc fluff, McSly, Mccready, Mec modifier, Mediumostrich, Meelar, Menezies C, Menhardtb, Mentifisto, Merovingian, Merphant, Mervyn, Metheone, Michael Devore, Michael Zimmermann, Michael93555, Michaelas10, Michiel.pruijssers, Microsoft mess u up, Mikalikali, Mike Rosoft, Mike dill, Mike14149393, Mikeroxbig, Mild Bill Hiccup, Minimac93, Minkykim101, MisfitToys, MisterHand, Misza13, Mking78, Moanzhu, Modulatum, Mohylek, Moink, Monedula, Monkeybait, Monkeychunk2, Monz, Mori Riyo, Morven, Mostsigbit, Motorrad-67, Mr Mulliner, Mr blobby eats carrots, MrOllie, MrRadioGuy, Mukerjee, Munkey Boy, Munkymu, Murcielago77, Murray Langton, Mwanner, My name, Myanw, Mydoorisopen, NJR ZA, Nagy, Namelessone7890123, Natalie Erin, Nathan123powell, NathanHurst, Nayvik, Nearfar, Neep, Neutrality, Newone, Nicksm, Nigel Ish, Nihiltres, Nilmerg, Niroht, Niteowlneils, Nixeagle, Nj8, Nlevitt, No Guru, NoIdeaNick, Nobleroyalties, NorwegianBlue, Nposs, Nsandre, NuclearWarfare, Obradovic Goran, Ocanter, Ocr31011, Ocsadanielsw@hotmail.com, Ocwaldron, Oda Mari, Ohnoitsjamie, Old Moonraker, Oldlaptop321, Oliver Pereira, Oliver202, OllieFury, Omegium, Omicronpersei8, Onceler, Onlandagain, Optichan, Orangutan, OverlordQ, Ovisoftblue, OwenX, Owned555, Oxymoron83, PARA19, PRA, PTSE, Pagrashtak, Pakaran, Pan Dan, Paranomia, ParisianBlade, Pashute, Pat Payne, Patrick, Paul W, Pawebster, Paxsimius, Pde, Pedant, Pediainsight, Pedrojpinto, Pengo, Pentasyllabic, Pentrant, PeopleString, Peppage, Pepperpiggle, Peregrine981, Peter Horn, Pgan002, Pgk, Phantombantam, Pharaoh of the Wizards, Pharos, Phidauex, Philip Baird Shearer, Philip Trueman, PhilipMW, Philippe, Phlebas, Piano non troppo, PierreAbbat, Pigsonthewing, Pilotguy, Pinkadelica, Pip2andahalf, Pizza1512, Placeholder account, Plek, Plymouths, Poe Joe, Pokemonsc1, Poorleno, Popageorgio, Poppy Joe, Porlob, Possum, Postdlf, Ppjr, Pratheepps, Preslethe, Prestonjb, Priceda, ProfDEH, PrometheusX303, Proofreader77, Proogs, Prospect77, PseudoSudo, Pyroclastic, Pyromaniac999, Pyromaniac9999, Quadell, QuadrivialMind, QuantumEleven, Quintote, Qwfp, Qxz, R27182818, R4ph, RCC1994, RCMoeur, RJaguar3, RMFan1, RMHED, Racker1, Radagast, RadicalBender, Raelx, Rahzel, RainbowOfLight, Rajeshd, Rama, Rami R, Rathfelder, Raul654, Raven42, Ravensfire, Ravidreams, RayKiddy, RazielZero, Rcc105, Rdsmith4, RealGrouchy, Rebroad, Reconsider the static, RedHillian, Redjar, Reflex Reaction, Rei, Reinthal, Res2216firestar, RexNL, ReyBrujo, Rgoodermote, Rholton, Riana, Rich257, Richard D. LeCour, Richfife, Rjanag, Rjd0060, Rjwilmsi, Rnelson, Roadahead, Robert Merkel, RobertG, Robertmh, Robin klein, Robinh, Robwingfield, Roger Irwin, Roguegeek, Romanceor, Rory096, RoyBoy, Rracecarr, Rsduhamel, Runner5k, Ryulong, S h i v a (Visnu), SCEhardt, SJP, SPUI, Sandahl, Sandius, Sandstein, SandyGeorgia, Sannse, Sasquatch, Saxophobia, Sceptre, Scetoaux, ScharfCustoms, SchfiftyThree, Sciencenerd69, Scott881994, Sdornan, SeanMack, Sebastian Henckel, Seegoon, Severo, Sf, Sfahey, Shadowjams, Sheldon Brown, Sheogorath, Sherool, Sherwoodseo, Shirulashem, Shredder000, Sidasta, Sietse Snel, Sigma 7, Signalhead, SimsFan, Sirex98, Siroxo, Sitethief, Sjakkalle, Sjorford, Sk8punk3d288, Skankkyeagle, Skatings mad, Skewwkey, Skier Dude, Skulliebones, Slakr, Slypolarmaniac56, Slysplace, Smack, Smalljim, Smerdis, Snigbrook, Snoyes, Soliloquial, Solipsist, Soneill83, Sonjaaa, Spalding, Spaomark, Spewin, Spider Pig, Spider Pig, does whatever a Spider Pig does, Spitfire, Springnuts, Staffwaterboy, Stannered, Statkit1, Steelfan36, Stephan Schulz, Stephen Gilbert, Stephenb, Stephengw, Steve125, SteveLoughran, Steven Weston, Steven Zhang, Stinky peterson, Struthious Bandersnatch, Stwalkerster, Suffusion of Yellow, SupaStarGirl, Super-Magician, SuperDude115, Superguy545, Superlaharl, Superm401, Surestick, Svedman, Switisweti, Synchronism, Sześćsetsześćdziesiątsześć, Sändaren, T0mmy12, THEN WHO WAS PHONE?, Tabletop, Tagishsimon, Taloooo, Tangotango, Tannerg99, Tariqabjotu, Tarquin, Taxman, Tayler1995, Tdbartz, Tedernst, TehRabbitt, Tempodivalse, Tensiomyography, Teslatesla, TestUser001, Textangel, Thaurisil, The 74.237.122.122, The Rambling Man, The S Man fo sho, The Thing That Should Not Be, The kid nigga, The wub, TheChrisD, TheDirtyTrorb, TheJC, Thecyberwasp, Theleftorium, Therightone, Thewallowmaker, Thewalrus, Thewayforward, Thincat, Thingg, Thomas Blomberg, Thomas-the-Rhymer, Three-quarter-ten, Thryduulf, Tide rolls, TigerShark, Tigersins88, Tim Starling, Tim1988, Timotab, Tiptoety, Tjunier, Tnasty08, Tobermcgee, Tombundy, Tony1, Tree Biting Conspiracy, Tregoweth, Trekphiler, Trent, Trevor MacInnis, Trevorparsons, Trilobite, Trimmer595, Trollderella, TrouserPress, Twilight Realm, Twinxor, Twiryan000, TwoDragons, Twooars, TylerJarHead, UFREE, Una Smith, UnitedStatesian, Until It Sleeps, Uriah923, Urod, User name one, Vald, Valentinian, Varmaa, Vclaw, VeloGuru, Ventusa, Verne Equinox, Vgedris, Viaciclante, Vicarious, Viperpepa, Viran, Virtualphtn, Visor, Vortexrealm, VoxLuna, Vreemdst, Vums, WJBscribe, WadeSimMiser, Waggers, Warfare utf, Wavelength, Wayward, Wbdrenner, Wesphys, Westmare, Weyes, Whereisscout, Wik, WikiDan61, WikipedianMarlith, Will 125, Willemkanon, William Ortiz, Willking1979, Wimt, Winkin, Wisdom man, Wknight94, Woodsalex, Woohookitty, Wprlh, Wtmitchell, Wuhwuzdat, Wwoods, X!, Xela, Xezbeth, Xxfelixxx, Yanksbuff, Yeeshenhao, Yekollu, Yoboybobbanana, Zadcat, Zanaq, Zapptastic, Zastava27, Zeeexsixare, Zhang He, Zigdon, ZmiLa, Zorzo Mirco, Zouavman Le Zouave, Ztrawhcs, Александър, 2029 anonymous edits

Freewheel *Source*: http://en.wikipedia.org/w/index.php?title=Freewheel *Contributors*: AHands, Alecjov, Beetstra, Bigdumbdinosaur, Biscuittin, Borb, Carpetburn, Christopherlin, Craig Pemberton, Dhodges, Diocito, DougieQ, Fratrep, Gregorydavid, Gzuckier, Heron, Honeyberry, Hooperbloob, Hugo999, JasperJ, Jengod, Keithonearth, Kjkolb, Leonard G., Maximus Rex, Mr Larrington, Mulad, Nlevitt, Nono64, Old Moonraker, Petershen1984, Ptr ru, Robofish, Rustemas, Scott Paeth, Segv11, Sfoskett, Shalom Yechiel, Six.oh.six, Sonett72, TarenGarond, Typ932, UninvitedCompany, Unknown Unknowns, User A1, Vincentyin, 49 anonymous edits

Sprocket *Source*: http://en.wikipedia.org/w/index.php?title=Sprocket *Contributors*: Ace Class Shadow, Acroterion, Af648, Alansohn, Alkiede, AndrewDressel, Atlant, Balabiot, Biscuittin, Bongwarrior, Capricorn42, Christopherlin, Claudelepoisson, Driwashsolutions, Ejay999, Enginerare, Ephraim33, Erik9, Erth64net, Faradayplank, Girolamo Savonarola, Gzuckier, G®iffen, Hooperbloob, Ian Dunster, Isilanes, Japanese Searobin, Jpc4031, Klayonic, Leonard G., MarkMLl, Martin451, Megapixie, MoCellMan, Murray Langton, Necessary Evil, NipponBill, Persian Poet Gal, Peter Horn, Ppe42, Retodon8, Savethemooses, ShakataGaNai, SpuriousQ, Stef4854, Tom harrison, Wereon, WikiLaurent, Ypacaraí, Zeimusu, 66 anonymous edits

Velodrome *Source*: http://en.wikipedia.org/w/index.php?title=Velodrome *Contributors*: *drowned*, 66FixedGear, 93JC, AMcDonald, Adam.J.W.C., Afinebalance, Alex0274, Amnesiak, Andycjp, Angusmclellan, AnnaFrance, Auntof6, Bedir26, Benstown, Black Falcon, Bornintheguz, Bryan Derksen, CDN99, Christopherlin, Conversion script, Covergaard, Crusoe8181, Dale Arnett, Danpat, Deon Steyn, Ds13, DuncanHill, Dysprosio, Eamonnca1, Editore99, Elkman, Elwood90, Emanuelbri, Eyal0, FigBug, Fixedgearfever, Flying Saucer, Fmiddleton, Fram, Gentgeen, Gryffindor, Gus.thomson, Headleyp, HendrikMalan, Hustvedt, Iblee, Icairns, Itsonlybarney, Jasmerb, Jdm, Jean.artegui, Jeff3000, Jeffhopkinsdlv, Jim856796, Jknotzke, Jmount, Jurjen, Kanabekobaton, KnowledgeOfSelf, Laurentseries, Lectonar, Lephilippe, Les woodland, Liftarn, Lightmouse, M.nelson, Mais oui!, Mandarax, Manos, Martarius, Mattbr, Merkin77, Mindfrieze, Mk32, Mr Larrington, Nadovich, Neelix, Neutralika, Nicholsr, Nklatt, Nopira, Nuttah, Oliver Han, Osomec, Owain, Paul W, Prolog, Qwfp, RedWolf, RingtailedFox, Rjwilmsi, Rogerzilla, RonGroth, Saint Midge, Schadler, Seventex, Severo, Shadow ump, Shell Kinney, Silenttxa, Speedeep, Stephen, Stephensuleeman, Stuntbaby, Surge79uwf, T3h 63n, Tense, Terrapin, Themfromspace, Thewalrus, Thoglette, To old, Tradnor, Traslaen, TwoOneTwo, Ukexpat, Underpants, Unyoyega, User2004, Versus22, WWC, Wa2go, Wavelength, Wfisher, Wikinewportwiki, Willy turner, Windynut, Zerge, Zootalures, 269 anonymous edits

Track bicycle *Source*: http://en.wikipedia.org/w/index.php?title=Track_bicycle *Contributors*: 842U, Adam.J.W.C., Andrew c, AndrewDressel, Binkitybonk, BitQuirky, Buffalo Bill, Cacophony, Christopher Parham, Christopherlin, Deon Steyn, Escape Orbit, Gdr, Gogo Dodo, Goworldgo, Herr Kriss, Hometack, Julius.kusuma, LDHan, Les woodland, Mattoondah, Mrmike23, Nixeagle, O1ive, Oscar, Phlebas, Rogerzilla, SCEhardt, Sakurambo, Severo, Smalljim, Thewalrus, Unregistered.coward, User name one, Watolson, Zeraien, 49 anonymous edits

Bicycle messenger *Source*: http://en.wikipedia.org/w/index.php?title=Bicycle_messenger *Contributors*: Agentbla, Ahalenia, Ashley Pomeroy, Beetstra, Bikecourier, BitQuirky, Brob, Buffalo Bill, COMPFUNK2, Cacophony, Carlosbenjamin, Cat Parade, Chris Bainbridge, Christopherlin, Colonies Chris, Commander Keane, Daysleeper47, Daytona2, Ds13, Dubaduba, Dysepsion, Dysprosia, ESkog, Emblemcycling, Erich168, Exander, Ferkelparade, Fkuehne, FrFintonStack, Frank duff, Future Perfect at Sunrise, Gdr, Glennwells, Gorgan almighty, HRpufnstuf, Hede2000, Ifny, Impacttrauma, Intensezkramer1993, JForget, JTN, Jamesywilliams, Jdorney, Jensentt, Julius.kusuma, Kchishol1970, Keithonearth, LDHan, LX, Lancekagar, Maelnuneb, Messville, Mini-Geek, Mipadi, Mrmike23, Mulad, Mythdon, NathanSmith, Neofelis Nebulosa, NormanBrown, Padraic, Pearle, Powerpedaler, ProfDEH, Rjwilmsi, Rogerzilla, S Chapin, SCEhardt, SchuminWeb, Severo, SidP, South Fulcrum, Stereotonic24, Strange but untrue, TheEgyptian, Thewalrus, Tillman, Trouts!, Velib, Viajero, Vicarious, Warreed, Wheresthebrain, Whistlingcyclist, Wimt, Wisden17, WiseElben, Wisl, WorldDownInFire, Wozzeck, Wwagner, Xompanthy, Zepheus, 203 anonymous edits

Bicycle brake systems *Source*: http://en.wikipedia.org/w/index.php?title=Bicycle_brake_systems *Contributors*: 08-15, A7N8X, ABF, Acroterion, Adamrush, Aezram, Alai, Alansohn, Alexburke, Alfio, Alpha 4615, AndrewDressel, Anticipation of a New Lover's Arrival, The, Aredding, Arencvier, Arnero, Barticus88, Blehargh, CL, Cabinscooter, Cataclysm, Ccrrccrr, Chowbok, Chris the speller, ChrisCork, Chrisahn, Christopherlin, Circumspice, Ciumtt, Ctbolt, DaemonLee, Dav nufc, Defenestrate, Deleting Unnecessary Words, Derek Ross, Djfitz111223, Dmforcier, DougsTech, Drewish, Ds13, EC 1234, Ebear422, Edgepg, Egil, Elipongo, Emiliobole, Epbr123, Eric Forste, Esperant, G-Man, Gerrit, Goodnightmush, Goof777, Graham87, Grand Edgemaster, Gurch, Gzuckier, Hairy Dude, Harryboyles, Hartleymartin, Hellbus, Heron, Hooperbloob, IW.HG, Igneousaddict, Inferno, Lord of Penguins, Infrogmation, Ionicism, JTN, Jaberwocky6669, Jeffrey.Kleykamp, JimmB, Johnkenyon, Jsallen1, Justtysen, Kafziel, Kaluza, Keanu, Keithonearth, Kerotan, Kingpin13, Kirtai, Klaws, Kozuch, LDHan, LaoHan, Leonard G., Liamoliver, Lightmouse, Loren.wilton, Magnus, Matt Gies, Menchi, Mentifisto, Mfero, Michael Daly, Michael Snow, Michellerun, Mike Nagle, Mork the delayer, Neilgunton, NekoDaemon, Nopetro, Old Moonraker, Omicronpersei8, Parhamr, PeramWiki, Peter Horn, Pgan002, Pissant, Raelx, RedWolf, Redjar, RijilV, Rmhermen, RonDivine, Ronwalf, Rosenknospe, SCEhardt, Salsa Shark, Sameerkale, Severo, Sfahey, Skittle, SkonesMickLoud, Snoyes, Soliloquial, SpiderJon, Sprouty76, Stephan Leeds, Stevage, SteveLoughran, TFBaschnagel, Tarheel95, Tarquin, Tellyaddict, That Guy, From That Show!, Themightyquill, Thinboy00, Think outside the box, Tresiden, Truthforlife, Understandwisdom, User name one, Versus22, Vintei, Visor, Will.law, Wmahan, Ww2censor, YankeeDoodle14, ZacharyS, Zharradan.angelfire, 321 anonymous edits

List of bicycle parts *Source*: http://en.wikipedia.org/w/index.php?title=List_of_bicycle_parts *Contributors*: 16@r, 842U, A. B., AHands, Agent007bm, Alan Liefting, AndrewDressel, Andrewa, Bardsandwarriors, Beland, Betterusername, BigFatBuddha, Christopherlin, DannyWilde, Davetrendler, Ddakah, De Mattia, Dhodges, Drunt, Ds13, Egil, Fletcher, Fred Bradstadt, Globbet, Herr Kriss, Ironman1104, JIl, Johnbryanpeters, JzG, Kalmia, Keithonearth, Kelaidis, Kellen`, Kev0153, Khukri, Leew, Lexor, Lilx, Markus Kuhn, Matt Gies, Mboverload, Mhockey, Michael Shields, Natkeeran, Nyxos, Pengo, Prestonjb, Rvollmert, Silverfish, Skoosh, Sonett72, Statkit1, Themightyquill, Thewalrus, Timothy Cooper, Twopinkwheels, Wernher, Yourdailywiki, 31 anonymous edits

Track cycling *Source*: http://en.wikipedia.org/w/index.php?title=Track_cycling *Contributors*: 128.103.11.xxx, 61.9.128.xxx, AHands, Adamrice, Aerolin55, Aff123a, Andrwsc, Ary29, Basement12, Biglovinb, Black Falcon, Bluelion, Bmpercy, Brentosthehungry, Bristee44, Bryan Derksen, C$, Cbuckley, Cf343b1, Christopherlin, Claywoolam, CoachMcGuirk, Connor Gilbert, Conversion script, Cordless Larry, D cushman, Dale Arnett, Davecrosby uk, David Krysakowski, Deon Steyn, Djharrity, Dlenmn, Drewish, Ds13, DuncanHill, Ekeb, Elpoca, Everyguy, Frank800, Gdr, Gothmog.es, Gzuckier, Hike395, Hotlorp, Ironman1104, Jay Litman, Jean.artegui, Jeronimo, Jim.henderson, JohnnoShadbolt, Josuechan, Kaushik twin, Kvn.rmndrz, LDHan, Lee Daniel Crocker, Les woodland, Lightmouse, Logophile, Markko, Miaow Miaow, O1ive, Orrelly Man, PAWiki, Paul W, Phlebas, Pirker, PoccilScript, Qwfp, RedWolf, Road Wizard, Sasquatch, Setikites, Severo, Siroxo, Slovenec sem, Sschnelk, Stocksy, Tabletop, Tainter, Thaf, The Thing That Should Not Be, Thewalrus, Twas Now, W., Wik, Will.law, Woohookitty, Xurei, YUL89YYZ, 爆笑連合, 128 anonymous edits

Image Sources, Licenses and Contributors

Image:Fixed-gear-bike.jpg *Source*: http://en.wikipedia.org/w/index.php?title=File:Fixed-gear-bike.jpg *License*: Creative Commons Attribution-Sharealike 2.5 *Contributors*: User:AndyArmstrong

Image:Fixed gear cog.jpg *Source*: http://en.wikipedia.org/w/index.php?title=File:Fixed_gear_cog.jpg *License*: Public Domain *Contributors*: User:John Reaves

Image:Track sprocket tool by Bruce McAdam.jpg *Source*: http://en.wikipedia.org/w/index.php?title=File:Track_sprocket_tool_by_Bruce_McAdam.jpg *License*: Creative Commons Attribution-Sharealike 2.0 *Contributors*: Bruce McAdam from Reykjavik, Iceland

Image:Fixed-gear-cogs.jpg *Source*: http://en.wikipedia.org/w/index.php?title=File:Fixed-gear-cogs.jpg *License*: Creative Commons Attribution-Sharealike 2.5 *Contributors*: User:AndyArmstrong

File:Courier 547.JPG *Source*: http://en.wikipedia.org/w/index.php?title=File:Courier_547.JPG *License*: unknown *Contributors*: User:ProfDEH

Image:Rear dropout.JPG *Source*: http://en.wikipedia.org/w/index.php?title=File:Rear_dropout.JPG *License*: GNU Free Documentation License *Contributors*: Andrew Dressel

Image:Marin bike.jpg *Source*: http://en.wikipedia.org/w/index.php?title=File:Marin_bike.jpg *License*: unknown *Contributors*: User:Roberta F.

Image:Pedelec-wildwind.jpg *Source*: http://en.wikipedia.org/w/index.php?title=File:Pedelec-wildwind.jpg *License*: Public Domain *Contributors*: CyclePat, MichaelDiederich, Wst

Image:Orbea Ordu.JPG *Source*: http://en.wikipedia.org/w/index.php?title=File:Orbea_Ordu.JPG *License*: unknown *Contributors*: User:AndrewDressel

Image:Draisine or Laufmaschine, around 1820. Archetype of the Bicycle. Pic 01.jpg *Source*: http://en.wikipedia.org/w/index.php?title=File:Draisine_or_Laufmaschine,_around_1820._Archetype_of_the_Bicycle._Pic_01.jpg *License*: unknown *Contributors*: User:Gun Powder Ma

Image:Michauxjun.jpg *Source*: http://en.wikipedia.org/w/index.php?title=File:Michauxjun.jpg *License*: Public Domain *Contributors*: unknown

Image:McCall1869.jpg *Source*: http://en.wikipedia.org/w/index.php?title=File:McCall1869.jpg *License*: unknown *Contributors*: Lesseps

Image:Ordinary bicycle01.jpg *Source*: http://en.wikipedia.org/w/index.php?title=File:Ordinary_bicycle01.jpg *License*: GNU Free Documentation License *Contributors*: User:Nova, User:Nova

Image:BicyclePlymouth.jpg *Source*: http://en.wikipedia.org/w/index.php?title=File:BicyclePlymouth.jpg *License*: Public Domain *Contributors*: AHands, Chris 73

Image:BicyclesMilkChurnsKolkata gobeirne.jpg *Source*: http://en.wikipedia.org/w/index.php?title=File:BicyclesMilkChurnsKolkata_gobeirne.jpg *License*: unknown *Contributors*: User:gobeirne

Image:Working bicycle.jpg *Source*: http://en.wikipedia.org/w/index.php?title=File:Working_bicycle.jpg *License*: Creative Commons Attribution-Sharealike 2.0 *Contributors*: Salim Virji

Image:Half Wheeler - bike.JPG *Source*: http://en.wikipedia.org/w/index.php?title=File:Half_Wheeler_-_bike.JPG *License*: unknown *Contributors*: User:PRA

Image:BMX bicycle.JPG *Source*: http://en.wikipedia.org/w/index.php?title=File:BMX_bicycle.JPG *License*: GNU Free Documentation License *Contributors*: Tukka

Image:Tuftscriterium.jpg *Source*: http://en.wikipedia.org/w/index.php?title=File:Tuftscriterium.jpg *License*: GNU Free Documentation License *Contributors*: Furmanj, Nhgrrl

Image:RacingBicycle-non.JPG *Source*: http://en.wikipedia.org/w/index.php?title=File:RacingBicycle-non.JPG *License*: GNU Free Documentation License *Contributors*: User:Niteowlneils

Image:Corsa bacchetta.jpg *Source*: http://en.wikipedia.org/w/index.php?title=File:Corsa_bacchetta.jpg *License*: GNU Free Documentation License *Contributors*: Original uploader was Jeffssmith1 at en.wikipedia

Image:Bicycle diagram-en.svg *Source*: http://en.wikipedia.org/w/index.php?title=File:Bicycle_diagram-en.svg *License*: unknown *Contributors*: User:Al2

Image:Triumph Bicycle.JPG *Source*: http://en.wikipedia.org/w/index.php?title=File:Triumph_Bicycle.JPG *License*: GNU Free Documentation License *Contributors*: AndrewDressel

Image:Shimano xt rear derailleur.jpg *Source*: http://en.wikipedia.org/w/index.php?title=File:Shimano_xt_rear_derailleur.jpg *License*: GNU Free Documentation License *Contributors*: C. Corleis

Image:Dsb-1.jpg *Source*: http://en.wikipedia.org/w/index.php?title=File:Dsb-1.jpg *License*: Attribution *Contributors*: Original uploader was Jeff dean at en.wikipedia

Image:Profile-2000-Aerobars.jpg *Source*: http://en.wikipedia.org/w/index.php?title=File:Profile-2000-Aerobars.jpg *License*: Creative Commons Attribution 2.5 *Contributors*: Jim.henderson

Image:San marco selle1.jpg *Source*: http://en.wikipedia.org/w/index.php?title=File:San_marco_selle1.jpg *License*: unknown *Contributors*: User:Roberta F.

Image:Brake.agr-edit.jpg *Source*: http://en.wikipedia.org/w/index.php?title=File:Brake.agr-edit.jpg *License*: GNU Free Documentation License *Contributors*: User:ArnoldReinhold, User:Visor

Image:BrakeDiskVR.JPG *Source*: http://en.wikipedia.org/w/index.php?title=File:BrakeDiskVR.JPG *License*: unknown *Contributors*: User:StromBer

Image:Mountainbike.jpg *Source*: http://en.wikipedia.org/w/index.php?title=File:Mountainbike.jpg *License*: unknown *Contributors*: Badseed, Er Komandante, Erik Kok, Herbythyme, Themightyquill, 7 anonymous edits

Image:Reiserad-beladen.jpg *Source*: http://en.wikipedia.org/w/index.php?title=File:Reiserad-beladen.jpg *License*: GNU Free Documentation License *Contributors*: User:Marcela

Image:Puncture-repaire-kit.jpg *Source*: http://en.wikipedia.org/w/index.php?title=File:Puncture-repaire-kit.jpg *License*: Creative Commons Attribution-Sharealike 2.0 *Contributors*: User:Warden, User:Warden

Image:BikesInAmsterdam 2004 SeanMcClean.jpg *Source*: http://en.wikipedia.org/w/index.php?title=File:BikesInAmsterdam_2004_SeanMcClean.jpg *License*: GNU Free Documentation License *Contributors*: Ilse@, Liftarn, MarcinJLewandowski, Vincent Steenberg, Wst

Image:Estacio bicing bcn.jpg *Source*: http://en.wikipedia.org/w/index.php?title=File:Estacio_bicing_bcn.jpg *License*: unknown *Contributors*: User:Marcbel

Image:Woman with Bicycle 1890s.jpg *Source*: http://en.wikipedia.org/w/index.php?title=File:Woman_with_Bicycle_1890s.jpg *License*: Public Domain *Contributors*: 16@r, Mwanner

File:Cycliste à place d'Italie-Paris.jpg *Source*: http://en.wikipedia.org/w/index.php?title=File:Cycliste_à_place_d'Italie-Paris.jpg *License*: unknown *Contributors*: Benchaum, Moebiusuibeom-en, Romanceor, Themightyquill

Image:Columbia Bicycles 1886 Advertisement.svg *Source*: http://en.wikipedia.org/w/index.php?title=File:Columbia_Bicycles_1886_Advertisement.svg *License*: Public Domain *Contributors*: w:Pope Manufacturing CompanyPope Manufacturing Company

Image:Person mit fahrrad feb07.jpg *Source*: http://en.wikipedia.org/w/index.php?title=File:Person_mit_fahrrad_feb07.jpg *License*: unknown *Contributors*: Walter Hochauer http://www.hochauer.net

File:Bicycle diagram reflectors.jpg *Source*: http://en.wikipedia.org/w/index.php?title=File:Bicycle_diagram_reflectors.jpg *License*: Public Domain *Contributors*: user:Javier Carro

Image:Freewheel en.svg *Source*: http://en.wikipedia.org/w/index.php?title=File:Freewheel_en.svg *License*: unknown *Contributors*: User:Borb, User:Technoargia

Image:Sprocket16.png *Source*: http://en.wikipedia.org/w/index.php?title=File:Sprocket16.png *License*: GNU Free Documentation License *Contributors*: Jahobr, Zeimusu

Image:Chain.gif *Source*: http://en.wikipedia.org/w/index.php?title=File:Chain.gif *License*: Public Domain *Contributors*: Juiced lemon, Ma-Lik, Mschlindwein, Tano4595, Wst

Image:Leclerc p1040882.jpg *Source*: http://en.wikipedia.org/w/index.php?title=File:Leclerc_p1040882.jpg *License*: unknown *Contributors*: User:David.Monniaux

Image:1Dunc Gray Velodrome.jpg *Source*: http://en.wikipedia.org/w/index.php?title=File:1Dunc_Gray_Velodrome.jpg *License*: unknown *Contributors*: User:Adam.J.W.C.

Image:Velodrome racing.jpg *Source*: http://en.wikipedia.org/w/index.php?title=File:Velodrome_racing.jpg *License*: GNU Free Documentation License *Contributors*: Burts, Elch, Jmabel

Image:Velodrome track markings.jpg *Source*: http://en.wikipedia.org/w/index.php?title=File:Velodrome_track_markings.jpg *License*: GNU Free Documentation License *Contributors*: Jmabel, Liftarn, Pirker

Image:Cyfac track bike.jpg *Source*: http://en.wikipedia.org/w/index.php?title=File:Cyfac_track_bike.jpg *License*: unknown *Contributors*: Julius.kusuma, Miffyandfrends, Mrmike23, Oda Mari, 11 anonymous edits

Image:Bicyclemeasurements.svg *Source*: http://en.wikipedia.org/w/index.php?title=File:Bicyclemeasurements.svg *License*: Public Domain *Contributors*: User:Sakurambo

Image:Bicycle courier 552.JPG *Source*: http://en.wikipedia.org/w/index.php?title=File:Bicycle_courier_552.JPG *License*: unknown *Contributors*: User:ProfDEH

Image:BikeBoys.jpg *Source*: http://en.wikipedia.org/w/index.php?title=File:BikeBoys.jpg *License*: unknown *Contributors*: Future Perfect at Sunrise, SCEhardt, Tillman, 2 anonymous edits

Image:Bicycle Messenger, London, 2006.jpg *Source*: http://en.wikipedia.org/w/index.php?title=File:Bicycle_Messenger,_London,_2006.jpg *License*: Creative Commons Attribution-Sharealike 3.0 *Contributors*: Buffalo Bill

Image:Rachel Picard winner 02.jpg *Source*: http://en.wikipedia.org/w/index.php?title=File:Rachel_Picard_winner_02.jpg *License*: Creative Commons Attribution 2.0 *Contributors*: FlickrLickr, FlickreviewR, Lumijaguaari

Image:Bicycle brakes - animated.gif *Source*: http://en.wikipedia.org/w/index.php?title=File:Bicycle_brakes_-_animated.gif *License*: Public Domain *Contributors*: User:Parhamr

Image:Marian038.jpg *Source*: http://en.wikipedia.org/w/index.php?title=File:Marian038.jpg *License*: GNU Free Documentation License *Contributors*: A7N8X, Adamrush, Mattes, Monobi, SCEhardt, 2 anonymous edits

File:Rod brake.JPG *Source*: http://en.wikipedia.org/w/index.php?title=File:Rod_brake.JPG *License*: unknown *Contributors*: User:AndrewDressel

Image:Bicycle caliper brake highlighted.jpg *Source*: http://en.wikipedia.org/w/index.php?title=File:Bicycle_caliper_brake_highlighted.jpg *License*: unknown *Contributors*: User:Aezram, User:keithonearth

Image:Dual-pivot calliper brakes .JPG *Source*: http://en.wikipedia.org/w/index.php?title=File:Dual-pivot_calliper_brakes_.JPG *License*: GNU Free Documentation License *Contributors*: AndrewDressel

Image:Bicycle centre pull brakes.jpg *Source*: http://en.wikipedia.org/w/index.php?title=File:Bicycle_centre_pull_brakes.jpg *License*: unknown *Contributors*: user:keithonearth

Image:Cantilever brake.JPG *Source*: http://en.wikipedia.org/w/index.php?title=File:Cantilever_brake.JPG *License*: GNU Free Documentation License *Contributors*: AndrewDressel, 1 anonymous edits

Image:Linear pull bicycle brake highlighted.jpg *Source*: http://en.wikipedia.org/w/index.php?title=File:Linear_pull_bicycle_brake_highlighted.jpg *License*: unknown *Contributors*: User:ArnoldReinhold, User:Keithonearth, User:Visor

Image:Roller Cam Bicycle Brake Front crop.JPG *Source*: http://en.wikipedia.org/w/index.php?title=File:Roller_Cam_Bicycle_Brake_Front_crop.JPG *License*: unknown *Contributors*: User:DMForcier

Image:Campagnolo Delta Brake front.jpg *Source*: http://en.wikipedia.org/w/index.php?title=File:Campagnolo_Delta_Brake_front.jpg *License*: GNU Free Documentation License *Contributors*: Campy Only

Image:Rollenbremse01.jpg *Source*: http://en.wikipedia.org/w/index.php?title=File:Rollenbremse01.jpg *License*: GNU Free Documentation License *Contributors*: Adamrush, Bartekbas, Javier Carro, Kneiphof, Markus Schweiss, Pitert, Themightyquill

Image:Rücktrittbremse geschnitten.jpg *Source*: http://en.wikipedia.org/w/index.php?title=File:Rücktrittbremse_geschnitten.jpg *License*: GNU Free Documentation License *Contributors*: Grenavitar, Kneiphof, Pitert, Stahlkocher, 1 anonymous edits

Image:Shimano 105-5500 shifters.jpg *Source*: http://en.wikipedia.org/w/index.php?title=File:Shimano_105-5500_shifters.jpg *License*: GNU Free Documentation License *Contributors*: Jim.henderson, Thewalrus

File:Bicycle diagram-en.svg *Source*: http://en.wikipedia.org/w/index.php?title=File:Bicycle_diagram-en.svg *License*: unknown *Contributors*: User:Al2

Image:Track cycling 2005.jpg *Source*: http://en.wikipedia.org/w/index.php?title=File:Track_cycling_2005.jpg *License*: unknown *Contributors*: Biopresto, Conscious, Pitert

Scientific Publishing House

offers

free of charge publication

of current academic research papers, Bachelor´s Theses, Master's Theses, Dissertations or Scientific Monographs

If you have written a thesis which satisfies high content as well as formal demands, and you are interested in a remunerated publication of your work, please send an e-mail with some initial information about yourself and your work to *info@vdm-publishing-house.com.*

Our editorial office will get in touch with you shortly.

VDM Publishing House Ltd.
Meldrum Court 17.
Beau Bassin
Mauritius
www.vdm-publishing-house.com

GNU Free Documentation License Version 1.2, November 2002 Copyright (C) 2000,2001,2002 Free Software Foundation, Inc. 59 Temple Place, Suite 330, Boston, MA 02111-1307 USA Everyone is permitted to copy and distribute verbatim copies of this license document, but changing it is not allowed.

0. PREAMBLE

The purpose of this License is to make a manual, textbook, or other functional and useful document "free" in the sense of freedom: to assure everyone the effective freedom to copy and redistribute it, with or without modifying it, either commercially or noncommercially. Secondarily, this License preserves for the author and publisher a way to get credit for their work, while not being considered responsible for modifications made by others. This License is a kind of "copyleft", which means that derivative works of the document must themselves be free in the same sense. It complements the GNU General Public License, which is a copyleft license designed for free software. We have designed this License in order to use it for manuals for free software, because free software needs free documentation: a free program should come with manuals providing the same freedoms that the software does. But this License is not limited to software manuals; it can be used for any textual work, regardless of subject matter or whether it is published as a printed book. We recommend this License principally for works whose purpose is instruction or reference.

1. APPLICABILITY AND DEFINITIONS

This License applies to any manual or other work, in any medium, that contains a notice placed by the copyright holder saying it can be distributed under the terms of this License. Such a notice grants a world-wide, royalty-free license, unlimited in duration, to use that work under the conditions stated herein. The "Document", below, refers to any such manual or work. Any member of the public is a licensee, and is addressed as "you". You accept the license if you copy, modify or distribute the work in a way requiring permission under copyright law. A "Modified Version" of the Document means any work containing the Document or a portion of it, either copied verbatim, or with modifications and/or translated into another language. A "Secondary Section" is a named appendix or a front-matter section of the Document that deals exclusively with the relationship of the publishers or authors of the Document to the Document's overall subject (or to related matters) and contains nothing that could fall directly within that overall subject. (Thus, if the Document is in part a textbook of mathematics, a Secondary Section may not explain any mathematics.) The relationship could be a matter of historical connection with the subject or with related matters, or of legal, commercial, philosophical, ethical or political position regarding them. The "Invariant Sections" are certain Secondary Sections whose titles are designated, as being those of Invariant Sections, in the notice that says that the Document is released under this License. If a section does not fit the above definition of Secondary then it is not allowed to be designated as Invariant. The Document may contain zero Invariant Sections. If the Document does not identify any Invariant Sections then there are none. The "Cover Texts" are certain short passages of text that are listed, as Front-Cover Texts or Back-Cover Texts, in the notice that says that the Document is released under this License. A Front-Cover Text may be at most 5 words, and a Back-Cover Text may be at most 25 words. A "Transparent" copy of the Document means a machine-readable copy, represented in a format whose specification is available to the general public, that is suitable for revising the document straightforwardly with generic text editors or (for images composed of pixels) generic paint programs or (for drawings) some widely available drawing editor, and that is suitable for input to text formatters or for automatic translation to a variety of formats suitable for input to text formatters. A copy made in an otherwise Transparent file format whose markup, or absence of markup, has been arranged to thwart or discourage subsequent modification by readers is not Transparent. An image format is not Transparent if used for any substantial amount of text. A copy that is not "Transparent" is called "Opaque". Examples of suitable formats for Transparent copies include plain ASCII without markup, Texinfo input format, LaTeX input format, SGML or XML using a publicly available DTD, and standard-conforming simple HTML, PostScript or PDF designed for human modification. Examples of transparent image formats include PNG, XCF and JPG. Opaque formats include proprietary formats that can be read and edited only by proprietary word processors, SGML or XML for which the DTD and/or processing tools are not generally available, and the machine-generated HTML, PostScript or PDF produced by some word processors for output purposes only. The "Title Page" means, for a printed book, the title page itself, plus such following pages as are needed to hold, legibly, the material this License requires to appear in the title page. For works in formats which do not have any title page as such, "Title Page" means the text near the most prominent appearance of the work's title, preceding the beginning of the body of the text. A section "Entitled XYZ" means a named subunit of the Document whose title either is precisely XYZ or contains XYZ in parentheses following text that translates XYZ in another language. (Here XYZ stands for a specific section name mentioned below, such as "Acknowledgements", "Dedications", "Endorsements", or "History".) To "Preserve the Title" of such a section when you modify the Document means that it remains a section "Entitled XYZ" according to this definition. The Document may include Warranty Disclaimers next to the notice which states that this License applies to the Document. These Warranty Disclaimers are considered to be included by reference in this License, but only as regards disclaiming warranties: any other implication that these Warranty Disclaimers may have is void and has no effect on the meaning of this License.

2. VERBATIM COPYING

You may copy and distribute the Document in any medium, either commercially or noncommercially, provided that this License, the copyright notices, and the license notice saying this License applies to the Document are reproduced in all copies, and that you add no other conditions whatsoever to those of this License. You may not use technical measures to obstruct or control the reading or further copying of the copies you make or distribute. However, you may accept compensation in exchange for copies. If you distribute a large enough number of copies you must also follow the conditions in section 3. You may also lend copies, under the same conditions stated above, and you may publicly display copies.

3. COPYING IN QUANTITY

If you publish printed copies (or copies in media that commonly have printed covers) of the Document, numbering more than 100, and the Document's license notice requires Cover Texts, you must enclose the copies in covers that carry, clearly and legibly, all these Cover Texts: Front-Cover Texts on the front cover, and Back-Cover Texts on the back cover. Both covers must also clearly and legibly identify you as the publisher of these copies. The front cover must present the full title with all words of the title equally prominent and visible. You may add other material on the covers in addition. Copying with changes limited to the covers, as long as they preserve the title of the Document and satisfy these conditions, can be treated as verbatim copying in other respects. If the required texts for either cover are too voluminous to fit legibly, you should put the first ones listed (as many as fit reasonably) on the actual cover, and continue the rest onto adjacent pages. If you publish or distribute Opaque copies of the Document numbering more than 100, you must either include a machine-readable Transparent copy along with each Opaque copy, or state in or with each Opaque copy a computer-network location from which the general network-using public has access to download using public-standard network protocols a complete Transparent copy of the Document, free of added material. If you use the latter option, you must take reasonably prudent steps, when you begin distribution of Opaque copies in quantity, to ensure that this Transparent copy will remain thus accessible at the stated location until at least one year after the last time you distribute an Opaque copy (directly or through your agents or retailers) of that edition to the public. It is requested, but not required, that you contact the authors of the Document well before redistributing any large number of copies, to give them a chance to provide you with an updated version of the Document.

4. MODIFICATIONS

You may copy and distribute a Modified Version of the Document under the conditions of sections 2 and 3 above, provided that you release the Modified Version under precisely this License, with the Modified Version filling the role of the Document, thus licensing distribution and modification of the Modified Version to whoever possesses a copy of it. In addition, you must do these things in the Modified Version: A. Use in the Title Page (and on the covers, if any) a title distinct from that of the Document, and from those of previous versions (which should, if there were any, be listed in the History section of the Document). You may use the same title as a previous version if the original publisher of that version gives permission. B. List on the Title Page, as authors, one or more persons or entities responsible for authorship of the modifications in the Modified Version, together with at least five of the principal authors of the Document (all of its principal authors, if it has fewer than five), unless they release you from this requirement. C. State on the Title page the name of the publisher of the Modified Version, as the publisher. D. Preserve all the copyright notices of the Document. E. Add an appropriate copyright notice for your modifications adjacent to the other copyright notices. F. Include, immediately after the copyright notices, a license notice giving the public permission to use the Modified Version under the terms of this License, in the form shown in the Addendum below. G. Preserve in that license notice the full lists of Invariant Sections and required Cover Texts given in the Document's license notice. H. Include an unaltered copy of this License. I. Preserve the section Entitled "History", Preserve its Title, and add to it an item stating at least the title, year, new authors, and publisher of the Modified Version as given on the Title Page. If there is no section Entitled "History" in the Document, create one stating the title, year, authors, and publisher of the Document as given on its Title Page, then add an item describing the Modified Version as stated in the previous sentence. J. Preserve the network location, if any, given in the Document for public access to a Transparent copy of the Document, and likewise the network locations given in the Document for previous versions it was based on. These may be placed in the "History" section. You may omit a network location for a work that was published at least four years before the Document itself, or if the original publisher of the version it refers to gives permission. K. For any section Entitled "Acknowledgements" or "Dedications", Preserve the Title of the section, and preserve in the section all the substance and tone of each of the contributor acknowledgements and/or dedications given therein. L. Preserve all the Invariant Sections of the Document, unaltered in their text and in their titles. Section numbers or the equivalent are not considered part of the section titles. M. Delete any section Entitled "Endorsements". Such a section may not be included in the Modified Version. N. Do not retitle any existing section to be Entitled "Endorsements" or to conflict in title with any Invariant Section. O. Preserve any Warranty Disclaimers. If the Modified Version includes new front-matter sections or appendices that qualify as Secondary Sections and contain no material copied from the Document, you may at your option designate some or all of these sections as invariant. To do this, add their titles to the list of Invariant Sections in the Modified Version's license notice. These titles must be distinct from any other section titles. You may add a section Entitled "Endorsements", provided it contains nothing but endorsements of your Modified Version by various parties--for example, statements of peer review or that the text has been approved by an organization as the authoritative definition of a standard. You may add a passage of up to five words as a Front-Cover Text, and a passage of up to 25 words as a Back-Cover Text, to the end of the list of Cover Texts in the Modified Version. Only one passage of Front-Cover Text and one of Back-Cover Text may be added by (or through arrangements made by) any one entity. If the Document already includes a cover text for the same cover, previously added by you or by arrangement made by the same entity you are acting on behalf of, you may not add another; but
you may replace the old one, on explicit permission from
previous publisher that added the old one. The author(s) a
publisher(s) of the Document do not by this License g
permission to use their names for publicity for or to assert or im
endorsement of any Modified Version.

5. COMBINING DOCUMENTS

You may combine the Document with other documents releas
under this License, under the terms defined in section 4 above
modified versions, provided that you include in the combination
of the Invariant Sections of all of the original documer
unmodified, and list them all as Invariant Sections of yo
combined work in its license notice, and that you preserve all th
Warranty Disclaimers. The combined work need only contain c
copy of this License, and multiple identical Invariant Sections m
be replaced with a single copy. If there are multiple Invari
Sections with the same name but different contents, make
title of each such section unique by adding at the end of it,
parentheses, the name of the original author or publisher of t
section if known, or else a unique number. Make the sa
adjustment to the section titles in the list of Invariant Sections
the license notice of the combined work. In the combination, y
must combine any sections Entitled "History" in the vari
original documents, forming one section Entitled "Histo
likewise combine any sections Entitled "Acknowledgements", a
any sections Entitled "Dedications". You must delete all sectio
Entitled "Endorsements".

6. COLLECTIONS OF DOCUMENTS

You may make a collection consisting of the Document and ot
documents released under this License, and replace
individual copies of this License in the various documents wit
single copy that is included in the collection, provided that
follow the rules of this License for verbatim copying of each of
documents in all other respects. You may extract a sin
document from such a collection, and distribute it individu
under this License, provided you insert a copy of this License i
the extracted document, and follow this License in all ot
respects regarding verbatim copying of that document.

7. AGGREGATION WITH INDEPENDENT WORKS

A compilation of the Document or its derivatives with ot
separate and independent documents or works, in or on a volu
of a storage or distribution medium, is called an "aggregate" if
copyright resulting from the compilation is not used to limit
legal rights of the compilation's users beyond what the indivic
works permit. When the Document is included in an aggreg
this License does not apply to the other works in the aggreg
which are not themselves derivative works of the Document. If
Cover Text requirement of section 3 is applicable to these cop
of the Document, then if the Document is less than one half of
entire aggregate, the Document's Cover Texts may be placed
covers that bracket the Document within the aggregate, or
electronic equivalent of covers if the Document is in electr
form. Otherwise they must appear on printed covers that brac
the whole aggregate.

8. TRANSLATION

Translation is considered a kind of modification, so you r
distribute translations of the Document under the terms of sec
4. Replacing Invariant Sections with translations requires spe
permission from their copyright holders, but you may incl
translations of some or all Invariant Sections in addition to
original versions of these Invariant Sections. You may includ
translation of this License, and all the license notices in
Document, and any Warranty Disclaimers, provided that you a
include the original English version of this License and
original versions of those notices and disclaimers. In case
disagreement between the translation and the original versio
this License or a notice or disclaimer, the original version
prevail. If a section in the Document is Ent
"Acknowledgements", "Dedications", or "History", the requiren
(section 4) to Preserve its Title (section 1) will typically req
changing the actual title.

9. TERMINATION

You may not copy, modify, sublicense, or distribute
Document except as expressly provided for under this Lice
Any other attempt to copy, modify, sublicense or distribute
Document is void, and will automatically terminate your ri
under this License. However, parties who have received cop
or rights, from you under this License will not have their licer
terminated so long as such parties remain in full compliance.

10. FUTURE REVISIONS OF THIS LICENSE

The Free Software Foundation may publish new, rev
versions of the GNU Free Documentation License from tim
time. Such new versions will be similar in spirit to the pre
version, but may differ in detail to address new problems
concerns. See http://www.gnu.org/copyleft/. Each version of
License is given a distinguishing version number. If the Docur
specifies that a particular numbered version of this License
any later version" applies to it, you have the option of follo
the terms and conditions either of that specified version or of
later version that has been published (not as a draft) by the F
Software Foundation. If the Document does not specify a ve
number of this License, you may choose any version
published (not as a draft) by the Free Software Founda
ADDENDUM: How to use this License for your documents To
this License in a document you have written, include a copy o
License in the document and put the following copyright
license notices just after the title page: Copyright (c) YEAR Y
NAME. Permission is granted to copy, distribute and/or m
this document under the terms of the GNU Free Documenta
License, Version 1.2 or any later version published by the F
Software Foundation; with no Invariant Sections, no Front-C
Texts, and no Back-Cover Texts. A copy of the license is inclu
in the section entitled "GNU Free Documentation License". If
have Invariant Sections, Front-Cover Texts and Back-C
Texts, replace the "with...Texts." line with this: with the Inva
Sections being LIST THEIR TITLES, with the Front-Cover T
being LIST, and with the Back-Cover Texts being LIST. If
have Invariant Sections without Cover Texts, or some
combination of the three, merge those two alternatives to sui
situation. If your document contains nontrivial example
program code, we recommend releasing these example
parallel under your choice of free software license, such as
GNU General Public License, to permit their use in free softw

CPSIA information can be obtained at www.ICGtesting.com
Printed in the USA
LVOW041805040213
318566LV00008B/1275/P